Sebastian Jakobi

Modern Aspects of Classical Automata Theory

T0135806

SEBASTIAN JAKOBI

Modern Aspects of Classical Automata Theory

Finite Automata, Biautomata, and Lossy Compression

Gießener Dissertation
Fachbereich Mathematik und Informatik, Physik, Geographie
Justus-Liebig-Universität Gießen 2014
1. Gutachter: Prof. Dr. Markus Holzer
2. Gutachter: Prof. Dr. Martin Kutrib
Termin der Disputation: 22.12.2014

Bibliografische Information der Deutschen Nationalbibliothek

Die Deutsche Nationalbibliothek verzeichnet diese Publikation in der
Deutschen Nationalbibliografie; detaillierte bibliografische Daten sind
im Internet über http://dnb.d-nb.de abrufbar.

ISBN 978-3-8325-3944-3

Logos Verlag Berlin GmbH
Comeniushof, Gubener Str. 47,
10243 Berlin
Tel.: +49 (0)30 42 85 10 90
Fax: +49 (0)30 42 85 10 92
INTERNET: http://www.logos-verlag.de

Preface

THIS thesis is situated in the fields of automata theory and formal languages, focusing mainly on regular languages and different types of language representations. The family of regular languages is maybe the most intensively studied and best understood class of formal languages. This language class exhibits many desirable properties which makes it suitable for a wide area of applications. One of its benefits is that a regular language can be represented by several different descriptional models, among which finite automata and regular expressions seem most common. The representation of languages by means of finite automata has a very algorithmic, machine-oriented flavor that allows direct implementation in computer systems. On the other hand, regular expressions often allow a more intuitive way of describing regular languages. This makes it an important task to convert between different forms of language representations. Moreover, in order to keep the memory requirements for storing the representation low, one may want to compress the size of a language representation. Now two questions arise: how does the size of the representation of a language change, when changing the underlying descriptional model? And how efficiently, in terms of computation time and required memory, can a compression of the representation be accomplished? These are typical questions from the fields of descriptional and, respectively, computational complexity theory. Since the early beginnings of automata and formal language theory, these questions have been thoroughly studied and today the descriptional and computational complexity of important conversion and compression problems is well-known.

In this thesis, we study modern aspects of compression and conversions of regular language representations. The first main part presents methods for lossy compression of finite automata. Lossy compression allows to reduce the size of a language representation below the limits of classical compression methods. But in return one has to put up with deviations from the original language. Thus, the compressed representation is not fully equivalent to the original language, but is only equivalent with respect to some errors. Besides existing methods of lossy compression, where one has only limited control on the committed errors, we introduce a more flexible notion of language equivalence, that allows to specify exactly the set of tolerated errors. But we will see that this more direct control on the set of errors comes at the cost of an increased computational complexity of lossy compression problems.

The other main part of this thesis is devoted to the study of biautomata, which were recently introduced as a new descriptional model for regular languages. We consider a more general biautomaton model and study several structural properties that affect descriptional and computational complexity aspects of these automata. The relation between biautomata and other descriptional models for regular languages is analyzed in terms of descriptional complexity of conversions between these models. Although biautomata are in many ways similar to finite automata, we will experience some notable differences. Concerning compression of language representations, we show how classical methods for finite automata can successfully be applied to biautomata. However, we observe a drastic increase of the computational complexity of lossy compression for biautomata, compared to finite automata.

Collaborations

Many results from this thesis emerged from joint work with Markus Holzer and were presented at the *16th* and *18th International Conference on Developments in Language Theory* [49, 55], the *18th International Conference on Implementation and Application of Automata* [50], the *15th International Workshop on Descriptional Complexity of Formal Systems* [51], and the *5th Workshop on Non-Classical Models of Automata and Applications* [52]. Revised and extended journal versions of [49], [51], and [52] appeared in [53], [56], and [57], respectively.

Acknowledgments

I would like to thank everyone who directly or indirectly helped me on my last years' journey as a doctoral candidate. First of all, I thank my advisor Markus Holzer for his manifold support: by countless research sessions also aside from my thesis topics, by encouraging me to collaborate with other researchers, and by spurring me on when I lapsed into a trot. I am also grateful to him and to Martin Kutrib for sending and accompanying me on many inspiring conferences and research travels. Further thanks go to all my colleagues for always being great company and for electing me C.o.M. in April 2014. On my private side, I am deeply thankful to my girlfriend Nadine not only for proofreading large parts of my thesis, but even more for giving me time freedom and limits when I needed them in stressful and, respectively, very stressful times—and of course for now being my wife. Last but not least, I would like to thank my parents Achim and Erika, and my brother Jörg for encouraging me to (maybe temporarily) leave my prospective vocational paths.

Contents

List of Figures

List of Tables

Part I

Preamble

Chapter 1

Introduction

ONE of the central concepts in the theory of formal languages has its source in linguistics. In the 1950s Noam Chomsky sought for a manageable formalism to describe and study the English language and grammar. In the papers [23, 24] Chomsky defined different types of formal grammars. Roughly speaking, such grammars represent formal languages by providing rules on how to generate "correct" sentences, or words of the respective language. The grammar types proposed by Chomsky differ from another in gradually increasing restrictions on the complexity of the grammar rules. Grammars which underlie only weak restrictions bear greater generative capacity than grammars that have to obey stronger restrictions. Hence, this hierarchy of grammars induces a hierarchy of classes of formal languages, which is nowadays known as the *Chomsky hierarchy*. The largest class of languages in that hierarchy, the family of *recursively enumerable languages*, is formed by grammars of type 0, where arbitrary rules can be used. Languages generated by the more restricted type 1 grammars are called *context-sensitive languages*. Such grammars allow to specify contexts in which certain rules may be applied. Grammars of type 2, where the applicability of rules cannot be controlled by contexts, characterize the family of *context-free languages*. The lowest level of the Chomsky hierarchy is formed by the most restrictive grammars of type 3, which generate so called *regular languages*.

These language families can also be characterized by means of automata theory. In 1936 Alan M. Turing introduced an abstract model of computing devices which he called *automatic machines*, aiming at formalizing the notions of computable numbers, computable functions, and so on [120]. This automaton model became a keystone of computer science and in honor of its inventor is called *Turing machine*. Such machines can be used as recognition devices for formal languages. Chomsky showed that the languages recognized, or accepted by Turing machines are exactly the recursively enumerable languages, that is, the topmost level of the Chomsky hierarchy [24]. Moreover, also the other language classes from the Chomsky hierarchy can be characterized as languages that are recognized by specific automaton models. Just as the different grammar types from the Chomsky hierarchy can be obtained from one another by gradually stronger restrictions, the corresponding automaton models can be seen as restricted Turing machines. These machines can be either deterministic or nondeterministic—roughly speaking, a nondeterministic machine may "guess" at certain points how to continue its computation—and can access the input in a one-way or two-way mode: one-way machines may only read their input once from left to right, while two-way machines can move their reading head in both directions on the input. In general, a Turing machine is a two-way machine that may work deterministically or nondeterministically—both models characterize the same language family. What the different automaton models have in common is that they consist of a finite control unit and an additional storage. The control unit consists of a finite number of internal states, representing a finite amount of information the machine can memorize, and an instruction table that controls the machine's operation. The additional storage depends on the automaton type. In case of Turing machines, the storage is a working tape (scratch paper) of unbounded length. If we restrict the length of the working tape to be linearly bounded in the length of the input given to the machine, we obtain a *linear bounded automaton*, which was first introduced as a deterministic model by John R. Myhill [99]. Sige-Yuki Kuroda showed that the family of languages accepted by *nondeterministic* linear bounded automata coincides with the context-sensitive languages [78]. Today it is still an open question, whether the same language class is obtained by considering *deterministic* linear bounded automata instead. Restricting the usage of the working tape to a LIFO ("last in first out") tape, that is, a push-down storage, and by imposing a one-way input mode, we obtain *push-down automata*, which were independently proposed by Anthony G. Oettinger [101] and Marcel-Paul Schützenberger [111]. It was shown by Chomsky [25] and by R. James Evey [32] that nondeterministic push-down automata accept exactly the context-free languages. The fact that deterministic push-down automata are strictly weaker than nondeterministic push-down automata was discovered independently by Seymour Ginsburg and Sheila Greibach [38], and by Leonard H. Haines [44]. Finally, by restricting the automata to use *no* additional storage at all, we obtain *finite automata*, which characterize the class of regular languages, regardless whether they are deterministic, nondeterministic, one-way, or two-way. The relation between finite automata and the lowest level of the Chomsky Hierarchy was reported by Noam Chomsky and George A. Miller [26].

From this point of view, the finite automaton is the fundamental computational device that forms the basis of other automaton models. Therefore, finite automata and the hereby characterized family of regular languages have been (and still are) subject to extensive and intensive research. The development of the finite automaton model is rooted in quite different areas. Already in 1943 Warren S. McCulloch and Walter Pitts used such devices for modeling activity in nerve nets [90]. In the 1950s David A. Huffman [63, 64] and George H. Mealy [93] used finite automata for studying switching circuits. Similar machines were also studied by Edward F. Moore [96]. All these models were purely deterministic in the sense that every computation step results in a successor state, which is uniquely defined by the machine's current state and the consumed input symbol. In 1959 Michael O. Rabin and Dana Scott introduced *nondeterministic* finite automata in their seminal paper [105], for which they were later granted the Turing award in 1976. Instead of entering a single predetermined successor state, a nondeterministic automaton has the possibility to choose in its computation steps between several possible successor states. Although at a first glance these automata seem more powerful than deterministic machines, Rabin and Scott showed that every nondeterministic finite automaton can be converted into an equivalent deterministic finite automaton, this means, into a deterministic automaton that accepts the same language as the nondeterministic one. Also equivalence of one-way and two-way finite automata was shown in that paper.

In addition to deterministic or nondeterministic finite automata, and type 3 grammars, the regular languages can be characterized by several other means. For example, there are algebraic characterizations of regular languages by finite monoids—see for example Jean-Éric Pin's chapter on syntactic semigroups [104] in the Handbook of Formal Languages [107]—and in terms of monadic second order logic as found by J. Richard Büchi [15], Calvin C. Elgot [29], and Boris A. Trakhtenbrot [118]. Another prominent descriptional formalism, which gave the class of regular languages its name, are the *regular expressions*, introduced in 1956 by Stephen C. Kleene [73]. Maybe due to their similarity to arithmetic expressions, regular expressions seem to be the most understandable and intuitive way for humans to describe regular languages. On the other hand, the machine-oriented finite automaton model seems best suited for software implementations.

This makes it an important task to convert between different representations of a regular language, say for example to convert a regular expression for some language into a finite automaton accepting that language. Luckily, there are effective algorithms for performing such conversions. However, some of these conversions are more complicated than others, which can be observed by means of complexity theory, in particular in computational complexity and in descriptional complexity theory.

Descriptional complexity theory is the study of economics of language descriptions, that is, how different descriptive systems can be used to represent languages succinctly. Here the succinctness is measured by different notions of sizes for descriptions of languages. For example, if the language is described by a finite automaton, then one can take the number of states of this automaton as the size of the description. First results on the descriptional complexity of regular languages date back to the paper from Rabin and Scott [105], where they introduced nondeterministic finite automata. Their conversion from nondeterministic to deterministic finite automata implies the follow *upper bound*: if the nondeterministic automaton has n states, then an equivalent deterministic machine needs *at most* 2^n states. However, at first it was not clear whether this construction was just too complicated and one could always construct smaller deterministic automata, or whether there are languages where such an exponential size blow-up cannot be avoided. This question was independently settled by Oleg B. Lupanov [85], Frank R. Moore [97], and by Albert R. Meyer and Michael J. Fischer [94]. They proved a corresponding *lower bound* for this conversion, by presenting families of languages that are accepted by n-state nondeterministic finite automata, and where every deterministic machine needs *at least* 2^n states. This proposes that nondeterministic automata are better suited to represent languages succinctly. Concerning conversions between finite automata and regular expressions it is known that a regular expression can be converted into a nondeterministic finite automaton that has about the same size as the expression. This is shown, for example, by a construction from Ken Thompson [116], which was also used in the implementation of the UNIX tool grep. Another well-known conversion from regular expressions to nondeterministic finite automata is due to Viktor M. Glushkov [40]. The conversion from a regular expression to a deterministic finite automaton is more complicated again. An exponential upper bound for the increase of description size can be concluded, for example, by Thompson's conversion from regular expressions to nondeterministic finite automata, followed by Rabin's and Scott's conversion from nondeterministic to deterministic finite automata. A more direct conversion, which also may result in an exponential size blow-up, was given by Janusz A. Brzozowski [13]. Exponential upper and lower bounds for this conversion were presented by Ernst Leiss [81], however his size measure for regular expressions was rather unconventional. More recent lower bound results can be found, for example, in a paper by Keith Ellul, Bryan Krawetz, Jeffrey Shallit, and Ming-Wei Wang [30]. What is known for the converse direction, that is, from finite automata to regular expressions? A well-known conversion technique is the state elimination method proposed by Janusz A. Brzozowski and Edward J. McCluskey [14]. Again, this conversion only provides an exponential upper bound for the size increase. Early exponential lower bound results for the conversion from finite automata to regular expressions were given by Andrzej Ehrenfeucht and Paul Zeiger [28], and more recently by Hermann Gruber and Markus Holzer [42], and by Wouter Gelade and Frank Neven [37].

The above mentioned results on an exponential increase of the description size when changing the descriptive system seem bad news. But it can get even worse, if we represent a regular language by a descriptive model that is "too powerful," like for example by a push-down automaton. Recall that push-down automata characterize the class of context-free languages, and this is a strict superset of the family of regular languages. Therefore, every regular language can be accepted by a push-down automaton, but not every language accepted by a push-down automaton is regular. Already in 1971 Meyer and Fischer [94] showed the following: there is a collection of regular languages that can be accepted by relatively small push-down automata, but the sizes of finite automata accepting these languages grow so fast that they cannot be bounded by any computable function that takes as argument the sizes of the push-down automata. This phenomenon is called a *non-recursive trade-off* and often occurs between descriptive systems that do not characterize the same language classes. An overview on results and proof-schemes for non-recursive trade-offs was given by Martin Kutrib [79].

Instead of a pessimistic view on the size blow-up that may occur when changing the descriptional system, one can obtain an optimistic view when considering the reverse conversion. If we are given a large deterministic finite automaton for some language, it may be possible to find a nondeterministic machine for that language, which is much smaller than the deterministic machine. In this way we could *compress* the representation of the language down to logarithmic size. Even more, if we have a non-recursive trade-off, then the savings in the description size could be arbitrary. However, we must not be too enthusiastic: although there may exist such compressed representations, it can be very hard, if not impossible, to compute them algorithmically. Another drawback of compressed representations is that it may be more complicated to extract information from them than it is to read this information from an uncompressed representation. This leads us to the before mentioned field of computational complexity theory.

Roughly speaking, computational complexity theory is devoted to the study of the question, which computational resources are sufficient and necessary for algorithms to solve specific tasks, or problems. The notion of an algorithm can be captured by the computational model of a Turing machine. Computational resources are then, for example, the running time or the memory used by the machine during its execution. Besides that, also the computational mode of the machine is taken into account, that is for example, whether the Turing machine is deterministic or nondeterministic. By restricting Turing machines to use only some specified time or space, or to work in a specific mode, only specific classes of problems can be solved. In this way, different *complexity classes* of problems are obtained. For example, the famous complexity class P consists of all problems that can be solved by deterministic Turing machines that have a running time which is polynomial in the size of their input. The complexity class NP is defined similar to P, but instead of deterministic Turing machines, nondeterministic machines are used. When thinking of real computers, problems in P can be solved in moderate computation time because there is a deterministic Turing machine, in other words a deterministic algorithm, that solves the problem in polynomial time. On the other hand, a nondeterministic Turing machine working in polynomial time describes a *nondeterministic* algorithm, which cannot directly be carried out by a computer that works deterministically. One possibility to implement a nondeterministic algorithm on a deterministic machine is to execute all possible computations that evolve from the nondeterministic choices in the algorithm. But this may lead to an exponential running time. Hence the two complexity classes P and NP are often seen as the border between practically tractable and intractable problems. However, it is not really clear whether they really mark this border, because *maybe* even the most complicated problems in NP can also be solved by deterministic Turing machines with a polynomial time bound, in other words, *maybe* P = NP holds. The identification of most complicated problems in NP, so called NP-complete problems, dates back to works by Stephen A. Cook [27], Richard M. Karp [72], and Leonid A. Levin [83]. The question whether P = NP or P ≠ NP—mostly the latter is expected to be true—is an important, unsolved problem in computer science. Moreover this problem is also on the list of the seven Millennium problems of the Clay Mathematics Institute, which offers a million dollar award for a solution to one of those problems.

Coming back to the above mentioned motivation of compressing the representation of a regular language, for example the following problem is of interest:

> given a finite automaton A, construct a finite automaton B that accepts the same language as A and that is as small as possible.

Instead of finite automata, also other descriptional models, like for example regular expressions, can be chosen. In these problems, the task is to compress the language representation A by *constructing* an equivalent but smaller representation B. Computational complexity theory mostly considers *decision problems*, which are problems allowing only "yes" and "no" answers. The decision version of the problem from above is the following *finite automata minimization problem*:

> given a finite automaton A and an integer n, decide whether there exists some finite automaton B that accepts the same language as A and that has at most n states.

The complexity of this problem depends on the automaton type. In case A and B are deterministic finite automata, the minimization problem is computationally easy, namely NL-complete, as shown by Sang Cho and Dung T. Huynh [22]—NL is a complexity class that is contained in P. In fact, for the problem of constructing a minimal deterministic finite automaton, many efficient algorithms are known, the most efficient of which was presented by John Hopcroft [61]. On the other hand, if nondeterministic automata are involved, then the problem becomes computationally intractable, namely complete for PSPACE, which is a complexity class that contains NP—see the paper of Tao Jiang and Bala Ravikumar [67]. This increase in computational complexity when changing from deterministic to nondeterministic finite automata is also observed in other relevant decision problems. For example the problem of deciding whether two finite automata are equivalent, that is, whether they accept the same language, is easy (NL-complete) for deterministic automata—see the paper of Cho and Huynh [22]—but complicated (PSPACE-complete) for nondeterministic automata, which follows from similar results on regular expressions, as to be found in a book of Alfred V. Aho, John E. Hopcroft, and Jeffrey D. Ullman [1] and a paper by Albert R. Meyer and Larry J. Stockmeyer [95].

Now let us come to more recent developments in language and automata theory. In 2009 Andrew Badr, Viliam Geffert, and Ian Shipman introduced the concept of hyper-minimization of finite automata [5]. The novelty of this concept is the following: classical minimization aims to compress a finite automaton while preserving the accepted language, whereas hyper-minimization allows the compressed automaton to accept a language that differs from the original language by a finite number of exceptions.[1] Thus, hyper-minimization can be seen as a kind of lossy compression of finite automata and it can allow more succinct representations, or rather approximations, of languages than classical minimization. It turned out that, similar to classical minimization, deterministic finite automata can be hyper-minimized in polynomial time. In fact, hyper-minimization algorithms with a similar running time as Hopcroft's minimization algorithm were developed independently by Paweł Gawrychowski and Artur Jeż [34], and by Markus Holzer and Andreas Maletti [59]. However, the exact computational complexity of hyper-minimization and related problems, in particular for nondeterministic finite automata, remained to be examined. One of the main parts of this thesis is concerned with the study of the complexity of such problems, as well as other types of lossy compression for finite automata.

A different line of research, that will be continued in this thesis, was initiated in 2012 by Ondřej Klíma and Libor Polák [74]. They introduced so called biautomata, a new automaton model that characterizes the class of regular languages. The motivation for this automaton model was to give new characterizations of important subclasses of the regular languages by restricting the structure of such biautomata. First studies on the descriptional complexity of conversions between biautomata and other descriptional models for regular languages were undertaken by Galina Jirásková and Ondřej Klíma [68]. The biautomaton model introduced by Klíma and Polák [74] is a deterministic machine and shows many similarities to deterministic finite automata. The present thesis also studies a nondeterministic and more general variant of biautomata, and carves out some interesting differences to finite automata.

The rest of this thesis is organized as follows. This introductory part continues in Chapter 2 by providing basic definitions and concepts used throughout this work. The body of the thesis then consists of two main parts.

Part II studies lossy compression of finite automata. First, different equivalence concepts from the literature are reviewed and a more general concept of language equivalence, which we call E-equivalence, is introduced: for a language E, we call two languages E-equivalent, if they differ only in words from E. These equivalence concepts can be seen as error profiles that specify, which errors are allowed in the course of lossy compression. Afterwards, the computational complexity of lossy compression and related problems for deterministic and nondeterministic finite automata is investigated. It turns out that most problems concerning nondeterministic automata are computationally intractable. For deterministic automata the complexity of the problems depends on the chosen error profile: while hyper-minimization of deterministic finite automata is computationally easy, the error profile of E-equivalence mostly yields

---

[1]The idea of allowing automata to disagree on a finite number of inputs was also used by Viliam Geffert, Carlo Mereghetti, and Giovanni Pighizzini [36] for a more efficient simulation of unary nondeterministic two-way automata.

intractable problems. Finally, it is shown how a prominent minimization algorithm of Brzozowski [12] can be turned into an algorithm for lossy compression of finite automata with respect to different error profiles.

In Part III the theory of biautomata is extended. A more general biautomaton model than introduced by Klíma and Polák [74] is defined and basic structural properties for biautomata are investigated. These properties affect various aspects of biautomata, such as the class of accepted languages, the descriptional complexity of conversion problems, and also the computational complexity of several decision problems. In particular, the absence of one those properties, which will later be called the diamond-property, induces non-recursive trade-offs and unsolvable minimization problems. Biautomata that satisfy important structural properties are related to classical descriptional models for regular languages in several ways. Concerning descriptional complexity, the relation between deterministic and nondeterministic biautomata resembles the situation for classical finite automata: a nondeterministic biautomaton may be exponentially more succinct than a deterministic one. Also the descriptional complexity of conversion problems between classical models and biautomata are studied. Another similarity to finite automata can be observed in terms of minimization procedures: classical minimization algorithms for finite automata can be successfully adapted to biautomata, even the rather untypical minimization algorithm of Brzozowski [12]. In fact, the computational complexity of minimizing biautomata is the same as for finite automata. The end of Part III takes up the idea of lossy compression from Part II again, by investigating the computational complexity of lossy compression problems for biautomata. Here a surprising difference between biautomata and finite automata shows up: in contrast to deterministic finite automata, where hyper-minimization is computationally easy, the hyper-minimization problem for deterministic biautomata is found to be computationally intractable.

Finally, Part IV concludes with a recapitulation of the main results from this thesis and points out possible directions for further research.

Chapter 2

Basic Concepts and Definitions

THIS chapter aims to explain the most relevant concepts used in this thesis, so that any interested reader with an appropriate background in computer science or mathematics will be able follow the studies herein. The first section fixes notations for formal words and languages, as well as operations on these objects. Then the family of regular languages is introduced in Section 2.2 along with the formalism of regular expressions for describing such languages. Section 2.3 presents important definitions and results on finite automata, devices which are used to recognize regular languages. Finally, some basics on computational complexity theory are discussed in Section 2.4. Of course, these introductions are kept rather short—except for the section on finite automata, which are central objects studied in this thesis. For more detailed treatments of the basics on theory of regular languages, formal languages and automata theory in general, and computational complexity theory, the reader is referred to Sheng Yu's chapter [126] in the Handbook of Formal Languages [107], the textbook by John E. Hopcroft, Rajeev Motwani, and Jeffrey D. Ullman [62], and the book of Christos H. Papadimitriou [102], respectively.

2.1   Words and Languages

The atomic objects in formal language theory, from which words and languages are built, are abstract *letters*, or *symbols*. A finite set of symbols is called an *alphabet*. A *word* over the alphabet Σ is a finite sequence of symbols from Σ. The *length* of a word w is the number of its symbols, and will be denoted by $|w|$. If $w = a_1 a_2 \ldots a_n$, with $a_1, a_2, \ldots, a_n \in \Sigma$, then $|w| = n$. Further, for a symbol $a \in \Sigma$ we denote by $|w|_a$ the number of occurrences of a in the word w. The *empty word* is the unique word of length zero, consisting of no symbols at all, and will be denoted by λ. The set of all words over the alphabet Σ is denoted by Σ^*, and a *language* over Σ is any subset of Σ^*. The *empty language* is denoted by $\emptyset$. For an integer n, we write $\Sigma^{<n}$ and $\Sigma^{\leq n}$ to denote the sets

of all words w over Σ with $|w| < n$ and, respectively, $|w| \leq n$. The sets $\Sigma^{>n}$ and $\Sigma^{\geq n}$ are similarly defined.

The *reversal* of a word $w \in \Sigma^*$ is the word $w^R \in \Sigma^*$ which is obtained by writing w backwards: if $w = a_1 a_2 \ldots a_{n-1} a_n$, then $w^R = a_n a_{n-1} \ldots a_2 a_1$. For the empty word we have $\lambda^R = \lambda$. Notice that for all words w, w_1 and w_2 we have $(w^R)^R = w$, and $(w_1 w_2)^R = w_2^R w_1^R$. The *concatenation* of two words w_1 and w_2 from Σ^* is the word $w_1 \cdot w_2 \in \Sigma^*$ built by juxtaposition of those two words: if $w_1 = a_1 a_2 \ldots a_n$ and $w_2 = b_1 b_2 \ldots b_m$, then $w_1 \cdot w_2 = a_1 a_2 \ldots a_n b_1 b_2 \ldots b_m$. For all words $w \in \Sigma^*$ we have $w = \lambda \cdot w = w \cdot \lambda$. Instead of $w_1 \cdot w_2$, we may often simply write $w_1 w_2$, omitting the concatenation dot. Formally, the set Σ^* can be seen as the monoid[1] generated by Σ, with the monoid operation $\cdot$ and the identity element λ. Formal powers of words are defined inductively via the concatenation operation: for a word $w \in \Sigma^*$ we define $w^0 = \lambda$, and $w^{i+1} = w \cdot w^i$, for all integers $i \geq 0$.

Now we come to operations on languages. Since languages are sets, we can apply basic set operations on them. We denote the *union* of L_1 and L_2 by $L_1 \cup L_2$ and their *intersection* by $L_1 \cap L_2$. The *difference* of the languages L_1 and L_2 is the set $L_1 \setminus L_2 = \{ w \in \Sigma^* \mid w \in L_1 \text{ and } w \notin L_2 \}$ and their *symmetric difference* is

$$L_1 \triangle L_2 = (L_1 \setminus L_2) \cup (L_2 \setminus L_1),$$

the set of all words that belong to exactly one of the two languages L_1 and L_2. Further, if L is a language over Σ, then its *complement* (with respect to Σ^*) is the language $\overline{L} = \Sigma^* \setminus L$, consisting of all words in Σ^* that do not belong to L. The previously defined operations reversal, concatenation, and formal powers of words can naturally be generalized to language operations as follows. Let L, L_1, and L_2 be languages over Σ. The *reversal* of L is the language $L^R = \{ w^R \mid w \in L \}$, and the *concatenation* of L_1 and L_2 is the language

$$L_1 \cdot L_2 = \{ w_1 \cdot w_2 \in \Sigma^* \mid w_1 \in L_1 \text{ and } w_2 \in L_2 \}.$$

Again, we may also write $L_1 L_2$ instead of $L_1 \cdot L_2$ to denote the concatenation of the languages L_1 and L_2. Formal powers of a language L are defined inductively by $L^0 = \{\lambda\}$, and $L^{i+1} = L \cdot L^i$, for all integers $i \geq 0$. The union of all formal powers of the language L is called *iteration*, or *(Kleene) star* of L, and is denoted by L^*, that is,

$$L^* = L^0 \cup L^1 \cup L^2 \cup \ldots = \bigcup_{i \geq 0} L^i.$$

By L^+ we denote the union $L^+ = \bigcup_{i \geq 1} L^i$ of all positive formal powers of L. Observe that the language L^* always contains the empty word, and that the Kleene star of the empty language $\emptyset$ is the set $\emptyset^* = \{\lambda\}$.

---

[1] A *monoid* is a set M with an operation $\bullet \colon M \times M \to M$ such that (i) there is a (unique) element $e \in M$, called the *identity element*, with $e \bullet m = m \bullet e = m$, for all $m \in M$, and (ii) the operation $\bullet$ is associative, i.e. $(\ell \bullet m) \bullet n = \ell \bullet (m \bullet n)$, for all $\ell, m, n \in M$.

Next, let us define derivatives of languages. Let L be some language and w some word over the alphabet Σ. The *left derivative* of L by w is the language

$$w^{-1}L = \{\, v \in \Sigma^* \mid wv \in L \,\},$$

and similarly, the *right derivative* of L by w is $Lw^{-1} = \{\, v \in \Sigma^* \mid vw \in L \,\}$. For all words $u, w \in \Sigma^*$, and all languages $L \subseteq \Sigma^*$, the following equality holds:

$$u^{-1}(Lw^{-1}) = \{\, v \in \Sigma^* \mid uvw \in L \,\} = (u^{-1}L)w^{-1}.$$

Therefore, we simply write $u^{-1}Lw^{-1}$ instead of $u^{-1}(Lw^{-1})$, and call this set the *both-sided derivative* of L by u and w. Instead of the term derivative, we may also use the name *quotient*.

Finally, we define a specific relation on words, which is induced by a language $L \subseteq \Sigma^*$. The *syntactic congruence* of L is the relation $\equiv_L$, defined for all words $u, v \in \Sigma^*$ such that $u \equiv_L v$ means that for all words $x, y \in \Sigma^*$ we have $xuy \in L$ if and only if $xvy \in L$. One can see that $\equiv_L$ is an equivalence relation on Σ^*. Moreover, if $u \in x^{-1}Ly^{-1}$ and $u \equiv_L v$, then also $v \in x^{-1}Ly^{-1}$. Therefore, every derivative of a language L is the union of some equivalence classes of $\equiv_L$.

2.2 Regular Languages and Regular Expressions

In this section we describe an important and long studied class of languages, the family of regular languages. Regular languages represent the lowest level in the Chomsky hierarchy of formal languages [23, 24]. There are various different ways of characterizing this language family; one approach is to say that regular languages are exactly those languages which can be obtained from the underlying alphabet symbols by finitely many applications of the operations concatenation, Kleene star, and union. This leads to the definition of regular expressions, introduced by Stephen C. Kleene [73], which are maybe the most convenient way for humans to describe a regular language.

A *regular expression* r over the alphabet Σ, and the *language* $L(r) \subseteq \Sigma^*$ *described* by r are inductively defined as follows:

- λ, $\emptyset$, and a, for symbols $a \in \Sigma$, are regular expressions, describing the languages $L(\lambda) = \{\lambda\}$, $L(\emptyset) = \emptyset$, and $L(a) = \{a\}$, respectively;

- if r, r_1, and r_2 are regular expressions, then r^*, $(r_1 \cdot r_2)$, and $(r_1 + r_2)$ are regular expressions, too, and their described languages are $L(r^*) = L(r)^*$, $L((r_1 \cdot r_2)) = L(r_1) \cdot L(r_2)$, and $L((r_1 + r_2)) = L(r_1) \cup L(r_2)$, respectively.

To keep the notation simple, it is conventional to give the star operator * highest precedence and let $\cdot$ precede $+$. Since concatenation and union are associative language operations, superfluous parentheses can be saved. Also, the concatenation operator $\cdot$ is often omitted, too, and we sometimes use the short-hand r^k, instead of writing $r \cdot r \cdot \ldots \cdot r$ (k times). Moreover, we use the notation r^+ for $r \cdot r^*$.

We say that $L \subseteq \Sigma^*$ is a *regular language* if there exists a regular expression r such that $L = L(r)$. The family of all regular languages is denoted by REG. When describing regular languages, we often identify the regular expression with the language it describes. For example, we may write $(b + ab^*a)^*$ instead of $L((b + ab^*a)^*)$ or $(\{b\} \cup (\{a\} \cdot \{b\}^* \cdot \{a\}))^*$ to describe the language of words over $\{a, b\}$, containing an even number of a symbols.

Example 2.1. Let $\Sigma = \{a, b, c\}$ and consider the regular expression

$$r = \left(\Big(\big((a + b) + c\big)^* \cdot (c \cdot c) \Big) \cdot \Big(\big((a + b) + c\big) \cdot \big((a + b) + c\big) \Big) \right)$$

over the alphabet $\Sigma = \{a, b, c\}$. As described above, this expression can be written in a more compact way as $r = (a + b + c)^* \cdot cc \cdot (a + b + c)^2$, and with the short-hand Σ for $(a + b + c)$, we can also write $r = \Sigma^* cc \Sigma^2$. The regular expression r describes the language $L(r)$ of all words in Σ^*, that consist of at least four symbols, and have symbol c at the third and fourth position from the right end. $\diamondsuit$

Next we want to discuss closure properties of the class of regular languages. A language class $\mathcal{L}$ is said to be *closed* under a language operation, if every application of that operation to languages from $\mathcal{L}$ yields a language from $\mathcal{L}$ again. Among the various language classes studied in formal language theory, the family of regular languages is probably the class that is closed under the most operations. Here we only want to mention a few operations, which are relevant in this work. Closure of REG under these operations is well known, see for example [62]. First of all, it is immediate from the definition of regular languages by regular expressions that the Kleene star, the concatenation, and the union of regular languages are regular, too. Therefore REG is closed under these operations. Another operation under which REG can easily seen to be closed is the reversal of languages: given a regular expression r, the language $L(r)^R$ can be described by reversing the expression r. Moreover REG is also closed under building derivatives of a language, as well as building the complement of a language (with respect to the underlying alphabet), which can easily be shown using an alternative characterization of regular languages by deterministic finite automata—see Section 2.3. Then, by De Morgan's law, closure of REG under intersection follows from closure under union and complementation.

We close this section with an observation on the number of derivatives of a regular language. This observation is based on important results from John R. Myhill [98] and Anil Nerode [100], which are also reported in the paper of Rabin and Scott [105]. These results show that a language $L \subseteq \Sigma^*$ is regular if and only if the number of equivalence classes of the syntactic congruence $\equiv_L$ on Σ^* is finite. Since derivatives of languages are unions of such equivalence classes, also the number of derivatives of a regular language must be finite.

Lemma 2.2. *Let $L \subseteq \Sigma^*$ be a regular language. Then the set*

$$\{\, u^{-1}Lv^{-1} \mid u, v \in \Sigma^* \,\}$$

of both-sided derivatives of L is finite. Moreover, also the number of left deriva-tives of L, as well as the number of right derivatives of L are finite. ☐

2.3 Finite Automata

The objects in the center of our study are finite automata. Such an automaton can be thought of as a machine that has an input tape on which some word is given and which is scanned by the machines reading head from left to right. The machine has some finite number of internal states and is controlled by a state transition function, which determines the next state of the machine, depending on the currently read input symbol and the internal state of the machine. After the machine has consumed its input it either accepts or rejects the input word, depending on whether it is in some distinguished accepting state or not.

Finite Automaton Model Definitions

A *nondeterministic finite automaton with nondeterministic initial state*, for short, an NNFA, is a quintuple $A = (Q, \Sigma, \delta, I, F)$ where

- Q is a finite set of *states*,

- Σ is the *input alphabet*,

- $\delta \colon Q \times \Sigma \to 2^Q$ is the *transition function*,

- $I \subseteq Q$ is the set of *initial states*, and

- $F \subseteq Q$ is the set of *final* or *accepting states*.

If for two not necessarily distinct states p and q from Q and an input $a \in \Sigma$ we have $q \in \delta(p, a)$, then we speak of a *transition* from p to q on input symbol a.

In case the NNFA $A = (Q, \Sigma, \delta, I, F)$ has a single initial state $q_0 \in Q$, that is, if $I = \{q_0\}$, then we simply speak of a *nondeterministic finite automaton* (NFA), and write $A = (Q, \Sigma, \delta, q_0, F)$, omitting the set braces for the initial state. If the automaton has a single initial state and further $|\delta(q, a)| = 1$ holds for all states $q \in Q$ and input symbols $a \in \Sigma$, then A is called a *deterministic finite automaton* (DFA). In this case the transition function δ can be thought of as a function $\delta \colon Q \times \Sigma \to Q$, mapping to states instead of sets of states. When the exact type of the automaton is not important, or clear from the context, we simply speak of a *finite automaton*.

The domain of the transition function δ of a finite automaton can naturally be extended to $Q \times \Sigma^*$ as follows. For all states $q \in Q$, symbols $a \in \Sigma$, and words $v \in \Sigma^*$ let

$$\delta(q, \lambda) = \{q\}, \quad \text{and} \quad \delta(q, av) = \bigcup_{p \in \delta(q, a)} \delta(p, v).$$

For deterministic finite automata, by interpretation of the transition function as a mapping to Q instead of 2^Q, this extension leads to a function $\delta\colon Q\times\Sigma^*\to Q$, with $\delta(q,\lambda)=q$, and $\delta(q,av)=\delta(\delta(q,a),v)$, for all states $q\in Q$, symbols $a\in\Sigma$, and words $v\in\Sigma^*$.

An NNFA $A=(Q,\Sigma,\delta,I,F)$ *accepts* a word $w\in\Sigma^*$ if there exists an initial state $q_0\in I$ and an accepting state $q_f\in F$, such that $q_f\in\delta(q_0,w)$. The *language accepted* by A is the language $L(A)=\{\,w\in\Sigma^*\mid A \text{ accepts } w\,\}$. These notions of acceptance naturally also apply to NFAs and DFAs. Notice that the language accepted by a DFA $A=(Q,\Sigma,\delta,q_0,F)$ can also be described by

$$L(A)=\{\,w\in\Sigma^*\mid \delta(q_0,w)\in F\,\},$$

when the transition function is interpreted as a function from $Q\times\Sigma^*$ to Q.

The *left language* of a state q in a finite automaton $A=(Q,\Sigma,\delta,I,F)$ is

$$L_{A,q}=\{\,w\in\Sigma^*\mid q\in\delta(q_0,w)\text{ for some }q_0\in I\,\},$$

the set of all inputs $w\in\Sigma^*$ that can take the automaton from an initial state to state q. Observe that $L_{A,q}$ is the language accepted by the finite automaton $A_q=(Q,\Sigma,\delta,I,\{q\})$ which is obtained from A by making state q the sole final state. A state $q\in Q$ is *reachable* in A if its left language $L_{A,q}$ is not empty. Moreover, for two states p and q of A we say that q is *reachable* from p, if $q\in\delta(p,w)$ for some word $w\in\Sigma^*$. The *right language* of a state q in A, denoted by $L_A(q)$, is defined as the set of input words that take the automaton from state q to some final state, that is,

$$L_A(q)=\{\,w\in\Sigma^*\mid \delta(q,w)\cap F\neq\emptyset\,\}.$$

In other words, the right language $L_A(q)$ is the language accepted by the finite automaton $_qA=(Q,\Sigma,\delta,\{q\},F)$, which results from A by making state q its single initial state. A state q of A is *useful* if its right language $L_A(q)$ is not empty. If the automaton A is clear from the context, we omit it in the index of the left and right languages, and write L_q instead of $L_{A,q}$, and $L(q)$ instead of $L_A(q)$.

Two finite automata A and B are called *equivalent*, denoted by $A\equiv B$, if they accept the same languages, that is, if $L(A)=L(B)$. A state p of a finite automaton A and a state q of a finite automaton B are *equivalent*, $p\equiv q$, if their right languages satisfy $L_A(p)=L_B(q)$—here the automata A and B may also be identical.

We illustrate above definitions in the following example, which also shows how finite automata can be represented by state transition diagrams.

Example 2.3. Consider the nondeterministic finite automaton $A=(Q,\Sigma,\delta,q_0,F)$ with state set $Q=\{q_0,q_1,q_2\}$, input alphabet $\Sigma=\{a,b\}$, initial state q_0, final state set $F=\{q_1,q_2\}$, and transition function δ defined as follows:

$$\delta(q_0,a)=\{q_0,q_1\},\qquad \delta(q_1,a)=\{q_2\},\qquad \delta(q_2,a)=\emptyset,$$
$$\delta(q_0,b)=\{q_0\},\qquad \delta(q_1,b)=\{q_2\},\qquad \delta(q_2,b)=\emptyset.$$

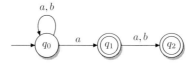

Figure 2.1: The nondeterministic finite automaton A over alphabet $\Sigma = \{a, b\}$ accepting the language $L(A) = \Sigma^* \cdot (a + ab)$.

The automaton A is depicted as a state transition graph in Figure 2.1. The vertices of the graph represent the states of the automaton; accepting states are drawn with a double circle, and initial states are marked by a (sourceless) incoming arrow. The state transitions are depicted by labeled arcs: whenever $q \in \delta(p, a)$ for some $p, q \in Q$ and $a \in \Sigma$, then there is an arc with label a from vertex p to vertex q.

The automaton A is *not* deterministic, because $|\delta(q_0, a)| \neq 1$. This can also be seen from Figure 2.1, because state q_0 has two outgoing arcs labeled with a. Also state q_2 violates the requirements of a deterministic machine, since no transitions are defined in this state.

Due to the loop in the initial state of A, we have $q_0 \in \delta(q_0, w)$, for all input words $w \in \Sigma^*$. Moreover, we have $q_1 \in \delta(q_0, w)$ if and only if w ends with symbol a, and $q_2 \in \delta(q_0, w)$ if and only if w ends with aa or ab. Hence the left language of the states of A are $L_{q_0} = (a + b)^*$, $L_{q_1} = (a + b)^*a$, and $L_{q_2} = (a + b)^*a(a + b)$, and the language accepted by A consists of all words ending with either a or ab, that is, $L(A) = (a + b)^*(a + ab)$. This is also the right language of the initial state q_0; the right languages of the other two states are $L(q_1) = \{\lambda, a, b\}$, and $L(q_2) = \{\lambda\}$. Since the right languages of all states in A are pairwise distinct, no two states of A are equivalent. $\diamondsuit$

Languages Accepted by Finite Automata

It is a well-known fact that every nondeterministic finite automaton (also with nondeterministic initial state) can be transformed into an equivalent deterministic finite automaton. This *determinization* can be achieved with the famous *power-set construction* [105] as follows. Let $A = (Q, \Sigma, \delta, I, F)$ be an NNFA. The *power-set automaton* of A is the deterministic finite automaton $\mathcal{P}(A) = (Q', \Sigma, \delta', q'_0, F')$, where the state set is $Q' = 2^Q$, the initial state is $q'_0 = I$, the set of final states is $F' = \{ P \in Q' \mid P \cap F \neq \emptyset \}$, and where the transition function $\delta' \colon Q' \times \Sigma \to Q'$ is defined for all states $P \in Q$ and input symbols $a \in \Sigma$ by

$$\delta'(P, a) = \bigcup_{p \in P} \delta(p, a).$$

The following result was shown by Rabin and Scott [105].

Theorem 2.4. *Let A be a nondeterministic finite automaton (with nondeterministic initial state). Then $L(\mathcal{P}(A)) = L(A)$.*

Let us give an example for the power-set construction.

Example 2.5. Let $A = (Q, \Sigma, \delta, q_0, F)$ be the NFA from Example 2.3, depicted in Figure 2.1. The equivalent power-set automaton $\mathcal{P}(A) = (Q', \Sigma, \delta', q_0', F')$ is depicted in Figure 2.2. The part on the right hand side of Figure 2.2 is grayed out because it is not reachable from the initial state $P_0 = \{q_0\}$ of the DFA $\mathcal{P}(A)$. These unreachable states can safely be omitted without changing the accepted language. Notice that the two states P_1 and P_2 of the automaton $\mathcal{P}(A)$ are equivalent, which can be seen as follows. Both states are accepting, so we have $\lambda \in L_{\mathcal{P}(A)}(P_1)$ and $\lambda \in L_{\mathcal{P}(A)}(P_2)$. Moreover, for both alphabet symbols a and b, state P_1 transitions to the same state as state P_2. Therefore, for all words $w \in \Sigma^+$ we have $w \in L_{\mathcal{P}(A)}(P_1)$ if and only if $w \in L_{\mathcal{P}(A)}(P_2)$. Thus, the right languages of P_1 and P_2 are equal, and we have $P_1 \equiv P_2$. $\diamond$

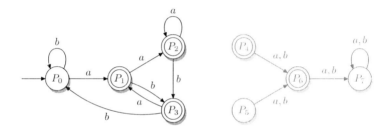

Figure 2.2: The power-set DFA $\mathcal{P}(A)$ constructed from the NFA A of Example 2.3. The grayed out part on the right formally belongs to the automaton $\mathcal{P}(A)$, but it is not reachable from the initial state of $\mathcal{P}(A)$. The state labels are abbreviations for the following subsets of $Q = \{q_0, q_1, q_2\}$: $P_0 = \{q_0\}$, $P_1 = \{q_0, q_1\}$, $P_2 = \{q_0, q_1, q_2\}$, $P_3 = \{q_0, q_2\}$, $P_4 = \{q_1, q_2\}$, $P_5 = \{q_1\}$, $P_6 = \{q_2\}$, and $P_7 = \emptyset$.

The power-set construction shows that the classes of languages accepted by the different types of above defined automata coincide. In fact, Kleene [73] showed that the languages that can be accepted by finite automata are exactly the regular languages.

Theorem 2.6. *A language L is regular if and only if there exists a finite automaton A with $L(A) = L$.*

In fact, one can effectively convert between different the representations of regular languages. Constructions for transforming a given regular expression into a finite automaton that accepts the language described by the expression, or *vice versa*, were given, for example, by Viktor M. Glushkov [40] and Robert McNaughton and Hisao Yamada [92]. In Section 8.2 in Part III we will see a transformation of regular expressions into so called *biautomata*, which are similar to finite automata.

Minimal Finite Automata

A basic measure for the size of a finite automaton is the number if its states. Sometimes also the number of transitions of the automaton is considered, but this number is polynomially related to the number of states: an n-state automaton over a k-letter alphabet has $n \cdot k$ transitions if it is deterministic, and at most $n^2 \cdot k$ transitions if it is nondeterministic. This motivates the following definition. A finite automaton A is *minimal* if there does not exists a finite automaton B of the same type, that is equivalent to A, but has fewer states than A. For example, a DFA A is minimal if there is no DFA B with fewer states, such that $A \equiv B$. However, for a minimal DFA A there may still be some equivalent *nondeterministic* finite automaton with fewer states—in this case A would still be a minimal DFA, but not a minimal NFA.

Although there are algorithms for constructing from a given nondeterministic finite automaton a minimal one, like the one by Tsunehiko Kameda and Peter Weiner [71], all known algorithms for minimizing NFAs in general are computationally very expensive. In fact it is known that the NFA minimization problem is computationally hard, see the works of Jiang and Ravikumar [67], and Andreas Malcher [86]—we will define more formally what hardness means in the upcoming Section 2.4.

Fortunately, for deterministic finite automata we are in a better situation. There are many efficient DFA minimization algorithms, the asymptotically fastest of which is Hopcroft's $O(n \log n)$ algorithm [61] which works by gradually refining a partition of the state set. There are also minimization algorithms that work the other way around, namely by coarsening a partition of the state set— a classification of such algorithms was given by Bruce W. Watson [123]. For many algorithms the idea, why they really construct a minimal automaton, is rather clear. An interesting algorithm, where this is not the case is Brzozowski's minimization algorithm [12]—we will study this algorithm and extensions of it in Chapter 5. Many minimization algorithms are based on structural characterizations of minimal deterministic finite automata. A very important result says that all minimal DFAs for a given regular language are identical, up to renaming of states [62]. Formally, this can be described as in the following theorem.

Theorem 2.7. *Let* $A = (Q, \Sigma, \delta, q_0, F)$ *and* $A' = (Q', \Sigma, \delta', q_0', F')$ *be two minimal deterministic finite automata with* $A \equiv A'$. *Then there exists a mapping* $h \colon Q \to Q'$ *that is bijective, and that satisfies the following conditions:*

1. $q \equiv h(q)$, for all $q \in Q$,

2. $h(q_0) = q_0'$,

3. $h(\delta(q, a)) = \delta'(h(q), a)$, for all $q \in Q$ and $a \in \Sigma$.

Notice that the first condition in particular says that h preserves acceptance of states, which means that $q \in F$ if and only if $h(q) \in F$. The third condition states that the mapping is compatible with taking transitions. This can also be generalized to words instead of symbols: $h(\delta(q, w)) = \delta'(h(q), w)$, for all $q \in Q$ and $w \in \Sigma^*$.

A useful description of the minimal DFA for a regular language in terms of its left derivatives is the canonical DFA [13]. Let L be a regular language over some alphabet Σ. The *canonical DFA* of L is $A_L = (Q_L, \Sigma, \delta_L, q_0^{(L)}, F_L)$ with the state set $Q_L = \{ u^{-1}L \mid u \in \Sigma^* \}$, initial state $q_0^{(L)} = L$, set of final states $F_L = \{ L' \in Q_L \mid \lambda \in L' \}$, and where the transition function is defined such that $\delta_L(L', a) = a^{-1}L'$, for all states $L' \in Q_L$ and symbols $a \in \Sigma$. Let us give an example for a canonical DFA.

Example 2.8. We want to describe the canonical DFA A_L for the regular language $L = \Sigma^*(a + ab)$ over the alphabet $\Sigma = \{a, b\}$—this is again the same language as accepted by the automaton A from Example 2.3 and its power-set automaton from Example 2.5. The initial state of A_L is the language L itself. The left derivatives of L by a and b are

$$a^{-1}L = L + \lambda + b, \qquad \text{and} \qquad b^{-1}L = L,$$

the derivatives of $a^{-1}L$ are

$$a^{-1}(a^{-1}L) = L + \lambda + b = a^{-1}L, \qquad \text{and} \qquad b^{-1}(a^{-1}L) = L + \lambda,$$

and the derivatives of $L + \lambda$ are the same as those of L. Therefore the DFA A_L has three states: $Q_L = \{ \Sigma^*(a+ab), \Sigma^*(a+ab)+\lambda, \Sigma^*(a+ab)+\lambda+b \}$, from which the last two are final states, and the first is the initial state. The transition function of A_L can be read from Figure 2.3. $\diamond$

Notice that automaton A_L from Example 2.8 has fewer states than the equivalent DFA $\mathcal{P}(A)$ from Example 2.5, even if we delete the unreachable states of $\mathcal{P}(A)$, so $\mathcal{P}(A)$ is not a minimal DFA. Remember that in the reachable part of $\mathcal{P}(A)$ the two states P_1 and P_2 are equivalent. In fact, the following characterization of minimal DFAs in terms of reachability and equivalence of states is well known, see for example [62].

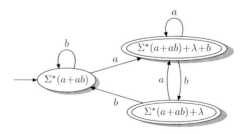

Figure 2.3: The canonical DFA A_L for the language $L = \Sigma^* \cdot (a + ab)$ over the alphabet $\Sigma = \{a, b\}$.

Theorem 2.9. *A deterministic finite automaton A is minimal if and only if all states of A are reachable and there is no pair of different but equivalent states in A.*

Theorem 2.9 leads to the following simple DFA minimization algorithm:

first delete all unreachable states, then merge all equivalent states.

Before we formally describe what is meant by state merging, we describe how equivalence of states can easily be determined by what is often referred to as the *table filling algorithm*. Let $A = (Q, \Sigma, \delta, q_0, F)$ be some DFA. First, we mark as *distinguishable* all pairs of states (p, q), where $p \in F$ and $q \notin F$, or where $p \notin F$ and $q \in F$. Then, as long as there is an unmarked state pair (p, q) and an input letter $a \in \Sigma$, such that the pair $(\delta(p, a), \delta(q, a))$ is marked, we also mark pair (p, q) as *distinguishable*. When no new marking can be done, two states p and q are equivalent if and only if the pair (p, q) is not marked.

Now let us formally define what is meant by state merging. Consider the DFA $A = (Q, \Sigma, \delta, q_0, F)$, and let $p, q \in Q$. The automaton obtained by *merging* state p to state q is the DFA $A' = (Q \setminus \{p\}, \Sigma, \delta', q_0', F \setminus \{p\})$, with

$$\delta'(r, a) = \begin{cases} q & \text{if } \delta(r, a) = p, \\ \delta(r, a) & \text{otherwise,} \end{cases} \quad \text{and} \quad q_0' = \begin{cases} q & \text{if } q_0 = p, \\ q_0 & \text{otherwise.} \end{cases}$$

If the states p and q of A are equivalent, then automaton A' is equivalent to A.

With Theorem 2.9 in mind, it is quite clear that this algorithm minimizes a given DFA. The following example illustrates the method of state merging.

Example 2.10. We start with the reachable part of the automaton $\mathcal{P}(A)$ from Example 2.5, which is depicted on the left in Figure 2.4. Recall that the states P_1 and P_2 are equivalent, so let us merge state P_1 to P_2. The obtained DFA is depicted on the right in Figure 2.4. Notice that this DFA is isomorphic to the canonical DFA A_L from Figure 2.3. Because this automaton has no pair of equivalent states, no further state merging can be done and the automaton is minimal. $\diamond$

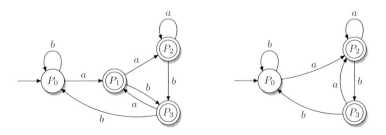

Figure 2.4: Left: the reachable part of the DFA $\mathcal{P}(A)$ from Example 2.5. Right: the DFA obtained from $\mathcal{P}(A)$ by merging state P_1 to state P_2.

2.4 Computational Complexity

In this section we present basics concepts from computational complexity theory that are used in this work. For more information, the reader is referred to the books of Michael R. Garey and David S. Johnson [33], Hopcroft et al. [62], and Papadimitriou [102]. We start with a very rough description of Turing machines, a computational model proposed by Alan M. Turing [120], which we use for two tasks: for accepting languages and for computing functions.

A *Turing machine* has a read-only *input tape*, one or more read- and writable *working tapes*, and if the machine computes a function, then it also has a write-only *output tape*. Tapes consist of linearly arranged tape cells, and in general the length of the tapes is not bounded. For each tape the machine uses a *read/write head* that can read and write one symbol at each computation step—the input head may only read symbols. Like a finite automaton, a Turing machine has a finite control consisting of a finite number of *states* and a *transition function*. If the machine is used to accept a language, then some states are distinguished *accepting states*. Depending on the currently read symbols on the input and working tapes, and the internal state of the machine, the transition function describes how to move its heads, and which symbols to write on the working and output tapes. Depending on whether this transition function allows only one or more than one possible next step, we speak of a *deterministic* or a *nondeterministic Turing machine*, respectively. In general, deterministic and nondeterministic Turing machines are equally powerful. Acceptance of languages from some input alphabet Σ is defined similar as for finite automata. If the content of the input tape is some word w from Σ^* and the Turing machine M ends its computation[2] in an accepting state, then we say that M *accepts* w, otherwise it *rejects* w. The language $L(M)$ consists of all words accepted by M. The *function computed* by a Turing machine M is the mapping $w \mapsto M(w)$, where $M(w)$ is the content of the output tape of M at the end of its computation on the input word w.

The computational problems investigated in this work are mainly decision problems.[3] Formally, a *decision problem* is a language L over some alphabet Σ. An *instance* of a problem is a word $w \in \Sigma^*$, and the task is to decide whether w belongs to L. If M is a Turing machine with $L(M) = L$, then we say that M *decides* L. However, we usually describe our problems more informally. As an example consider the *graph reachability problem*, which can be found in papers of Neil D. Jones [69] and Neil D. Jones, Y. Edmund Lien, and William T. Laaser [70], and which is formulated as follows:

> given a directed graph $G = (V, E)$ and two vertices $s, t \in V$, decide whether t is reachable from s.

---

[2] In this work we only need to consider Turing machines that halt on every input, that is, no computation may run indefinitely.

[3] In Section 4.4 we will also consider counting problems. Definitions of classes of counting problems can be found in that section.

Formally we would have to describe an encoding of graphs and vertices (for example over the binary alphabet $\{0,1\}$), and then define the problem as the set of all words $w \in \{0,1\}^*$, which are encodings of a graph G and vertices s and t, where t is reachable from s.

Now we can define classes of decision problems by restricting the computational resources of Turing machines. We first consider time bounded computation. For a function $f \colon \mathbb{N} \to \mathbb{N}$, the class $\mathsf{DTIME}(f(n))$ contains all problems which can be decided by some deterministic Turing machine, which halts after at most $f(n)$ computation steps, if given an input of length n. The class $\mathsf{NTIME}(f(n))$ is defined similarly, by using nondeterministic instead of deterministic Turing machines. The classes of all problem solvable by deterministic, and by nondeterministic Turing machines in polynomial time are

$$\mathsf{P} = \bigcup_{k \in \mathbb{N}} \mathsf{DTIME}(n^k) \qquad \text{and} \qquad \mathsf{NP} = \bigcup_{k \in \mathbb{N}} \mathsf{NTIME}(n^k).$$

Now let us define complexity classes of space bounded computation. For a function $f \colon \mathbb{N} \to \mathbb{N}$ we denote by $\mathsf{DSPACE}(f(n))$ the class of problems that are decidable by deterministic Turing machines which, if given an input of length n, use at most $f(n)$ tape cells on their working tapes. The class $\mathsf{NSPACE}(f(n))$ is the corresponding class of problems that are decidable by $f(n)$ space bounded nondeterministic Turing machines. In this work we consider logarithmic and polynomial space bounds: the classes of problems decided by logarithmic space bounded machines are

$$\mathsf{L} = \mathsf{DSPACE}(\log n) \qquad \text{and} \qquad \mathsf{NL} = \mathsf{NSPACE}(\log n),$$

and for polynomial space bounds we define

$$\mathsf{PSPACE} = \bigcup_{k \in \mathbb{N}} \mathsf{DSPACE}(n^k) \qquad \text{and} \qquad \mathsf{NPSPACE} = \bigcup_{k \in \mathbb{N}} \mathsf{NSPACE}(n^k).$$

For every complexity class C of decision problems we denote the set of problems the complements of which belong to C by $\mathsf{coC} = \{\, \overline{L} \mid L \in \mathsf{C} \,\}$. Since deterministic complexity classes are closed under complementation—just interchange accepting and non-accepting states—we know $\mathsf{P} = \mathsf{coP}$, $\mathsf{L} = \mathsf{coL}$, and $\mathsf{coPSPACE} = \mathsf{PSPACE}$. Moreover, by the seminal results independently obtained by Neil Immerman [65] and Róbert Szelepcsényi [113], it is known that also nondeterministic space bounded complexity classes are closed under complementation, that means, $\mathsf{NL} = \mathsf{coNL}$ and $\mathsf{NPSPACE} = \mathsf{coNPSPACE}$. However, it is still an open problem to prove or disprove $\mathsf{NP} = \mathsf{coNP}$—which is commonly believed not to hold. The following inclusions between above defined complexity classes are well known:

$$\mathsf{L} \subseteq \mathsf{NL} \subseteq \mathsf{P} \subseteq \mathsf{NP} \subseteq \mathsf{PSPACE} = \mathsf{NPSPACE},$$

where the equality of the classes PSPACE and $\mathsf{NPSPACE}$ follows from results by Walter J. Savitch [109]. Whether the inclusions are strict is yet unknown, but this is mostly expected to be the case.

Next we describe what makes a problem "typical" for a complexity class. The central concept here is the reduction of problems. A *reduction* from problem $L \subseteq \Sigma^*$ to problem $L' \subseteq \Gamma^*$ is a function $f\colon \Sigma^* \to \Gamma^*$ such that for all $w \in \Sigma^*$ we have

$$w \in L \text{ if and only if } f(w) \in L'.$$

This means that the reduction transforms an instance of L into an "equivalent" instance of problem L', so that we can solve L by solving L'. Hence, if the computation of the reduction is not "too expensive," then solving L cannot be much more difficult than solving L'. The expensiveness of a reduction is formalized as follows. We say that L is *polynomial time reducible* to L' if there is a reduction f from L to L' which can be computed by a deterministic polynomial time bounded Turing machine, and L is *logarithmic space reducible* to L' if the reduction can be computed by a deterministic logarithmic space bounded Turing machine. Every logarithmic space reduction is also a polynomial time reduction.

Now we can define "difficult" and "typical" problems: a problem L' is *hard* for a complexity class C, for short, C-*hard*, if every problem $L \in$ C is reducible to L'. The problem L' is C-*complete* if it is C-hard and $L' \in$ C. Notice that the notions of hardness and completeness depend on the type of reduction that is used. Generally, to obtain meaningful results, the reduction should not be too powerful with respect to the class of problems at hand: using polynomial time reductions, every problem is NL-hard because the Turing machine for the reduction can already solve all problems in NL. Therefore when speaking of NL-hardness we always assume logarithmic space reductions, while polynomial time reductions are sufficient for hardness in classes above P.

These concepts are used to classify the complexity of problems. Problems which are solvable in polynomial time, that is, which belong to P, or even to some class below, are said to be *computationally tractable*, and problems, which are known to be hard for some complexity class above P, are called *intractable*.

Part II

Lossy Compression of Finite Automata

Chapter 3

Error Profiles

ENERALLY, two automata can be considered equivalent if they show the same behaviour, where the term "behaviour" can have different meanings: in case of automata that are used to accept languages, the behaviour of an automaton can simply be defined as its accepted language—this leads to the equivalence relation ≡ from Section 2.3. If instead of accepting languages, the automata are used to produce output, like for example in Mealey or Moore machines [93, 96], then the behaviour can be defined via the relation between the machine's input and output. More sophisticated definitions of "behaviour" take into account the structure of automata and the sequence of transitions and states visited during computations, see for example the paper of Raymond T. Yeh [124]. Also for formal grammars, which generate words by successively applying specific rules, concepts of structural equivalence are considered, for example by Robert McNaughton [91] and Marvin C. Paull and Stephen H. Unger [103]. Another form of equivalence, used throughout different areas of computer science and other fields, is the concept of bisimulation—a thorough investigation of the origins of bisimulation is given by Davide Sangiorgi [108] and recent applications of bisimulation on finite automata can be found, for example, in the paper of Filippo Bonchi and Damien Pous [10].

In the following, we are mainly interested in the languages accepted by automata. Generally, the above mentioned concepts of behavioral equivalence imply equality of the accepted languages. But for some applications, automata may be considered "equally good" even if their languages do not exactly match. For example, a fundamental task in model checking—see the book of Christel Baier and Joost-Pieter Katoen [6]—is to test whether a given system model complies with some given specification. In order to test this, one checks whether the language that describes the behaviour of the system is *included* in the language described by the specification. A special case of language inclusion can also be found in the setting of cover automata, which were introduced by Cezar Câmpeanu, Nicolae Sântean, and Sheng Yu [17]. Such automata are used for concise representation of finite languages: here the automaton has to comply with all words from the original language, but it may also accept words that are longer than all words in the given finite language. Somehow complementary to cover automata are the notions of almost-equivalence and k-equivalence, introduced by Badr et al. [5] and Gawrychowski and Jeż [34], respectively: here languages may only differ on short words. We will later come back to the concepts of cover automata, almost-equivalence, and k-equivalence.

The equivalence concepts where the languages need not be equal can be interpreted as *error profiles* in the following sense. For a given automaton accepting some "original" language L, we consider all automata that are somehow equivalent to the given one as possible ersatz automata that accept "approximations" of L. The deviations between the original and the approximation language, that is, the words in the symmetric difference of the two languages, are the "errors"[1] committed by the ersatz automata. Because of their use in many applications, an important task is the minimization of automata, where one wishes to reduce the size of a given automaton, while preserving desired aspects of its behaviour. Classically, the reduced automaton is required to be language equivalent to the given one. If the automata need only be equivalent with respect to some error profile, then the minimization can be seen as a kind of lossy compression: by allowing deviations from the original language one may gain more succinct representations of languages.

The first part of this thesis is concerned with the study of lossy compression of finite automata. We mainly consider two error profiles which will be introduced in this chapter: almost-equivalence and E-equivalence. Basic results from the literature on almost-equivalence, which are relevant for our work, will be reviewed in the upcoming Section 3.1. Then in Section 3.2 the more general error profile of E-equivalence will be defined. This equivalence notion allows to explicitly define the set of allowed errors. Moreover, E-equivalence in some sense covers the notion of almost-equivalence, as well as other error profiles considered in the literature—this will shortly be discussed in Section 3.3.

The computational complexity of equivalence and minimality questions with respect to different error profiles will be studied in Chapter 4. Besides decision problems, we also consider counting problems for finite automata. We will see that problems related to the notion of almost-equivalence are of similar complexity as the corresponding problems for classical equivalence. On the other hand, when changing from almost-equivalence to E-equivalence, the complexity of some problems notably increases.

Chapter 5 develops algorithms for constructing deterministic finite automata that are minimal with respect to specific error profiles. These are obtained from Brzozowski's minimization algorithm [12] by computing additional information on E-equivalence in an intermediate stage of the algorithm. Using appropriate error sets for E-equivalence, algorithms for different error profiles are obtained.

---

[1]Errors or faults in automata were studied earlier, for example by Raymond Boute and Edward J. McCluskey [11]. There, attention was mainly paid to the detection of faults in sequential circuits and to the study of relations between such faults, rather than to relate languages or automata that are "similar" with respect to different errors.

3.1 Almost-Equivalence and Hyper-Minimality

The notions of almost-equivalence and hyper-minimality were introduced by Badr et al. [5]. Two languages L and L' are *almost-equivalent*, denoted by $L \sim L'$, if their symmetric difference $L \triangle L'$ is finite. As for classical equivalence ($\equiv$), this relation carries over to finite automata and states. Two finite automata A and A' are *almost-equivalent*, denoted by $A \sim A'$, if their accepted languages $L(A)$ and $L(A')$ are almost-equivalent. Similarly, two states q and q' of finite automata A and A', respectively, are called *almost-equivalent*, $q \sim q'$, if their right languages $L_A(q)$ and $L_{A'}(q')$ are almost-equivalent. It is known that almost-equivalence is an equivalence relation [5]. A deterministic (or nondeterministic) finite automaton A is *hyper-minimal*, if there is no deterministic (or nondeterministic, respectively) finite automaton A', with $A \sim A'$, that has fewer states than A.

A first hyper-minimization algorithm for DFAs was developed by Andrew Badr, Viliam Geffert, and Ian Shipman [5], with a running time of $O(n^3)$, where n is the number of states of the input DFA, assuming a constant size input alphabet. This was followed by an $O(n^2)$ time algorithm by Badr [4], before independently Gawrychowski and Jeż [34], and Holzer and Maletti [59] presented hyper-minimization algorithms with running time $O(n \log n)$. It is worth noting that this is also the running time of the fastest known algorithm for classical DFA minimization, which is Hopcroft's algorithm [61]. Andreas Maletti and Daniel Quernheim [88] described an $O(n^2)$ time algorithm which constructs, for a given DFA A, an almost-equivalent hyper-minimal DFA A' that commits the least number of errors, that is, where $|L(A) \triangle L(A')|$ is minimal, among all hyper-minimal DFAs that are almost-equivalent to A. In Section 5.3 we will describe another hyper-minimization algorithm based on Brzozowski's DFA minimization algorithm [12]. Hyper-minimization was also studied for other automaton models, like for example tree automata (see Jeż and Maletti [66]), weighted automata (see Maletti and Quernheim [89]), and Büchi automata (see Sven Schewe [110]).

In the following we take a closer look on the structure of almost-equivalent and hyper-minimal DFAs. At the end of the section we will then derive a bound for the maximum possible error length that can occur between almost-equivalent regular languages.

On the Structure of Almost-Equivalent DFAs

Central to the study of almost-equivalent DFAs are the preamble and kernel of an automaton. Let $A = (Q, \Sigma, \delta, q_0, F)$ be a DFA. A state $q \in Q$ is a *kernel state* if it is reachable from the initial state q_0 by an infinite number of inputs, and it is a *preamble state* if it is reachable from q_0 only by a finite number of inputs. In other words, if a state's left language is infinite, then it is a kernel state, otherwise it is a preamble state. The *preamble* of A is the set $\mathrm{Pre}(A)$ of all preamble states of A, and the *kernel* of A is the set $\mathrm{Ker}(A)$ of kernel states of A. We will often use the fact that the length of a word, that leads to a preamble state of A, is at most $|\mathrm{Pre}(A)| - 1$. This can be seen by a simple pumping argument. The following characterization of hyper-minimal DFAs, which is similar to the characterization of minimal DFAs as in Section 2.3, was presented by Badr et al. [5].

Theorem 3.1. *A deterministic finite automaton $A = (Q, \Sigma, \delta, q_0, F)$ is hyper-minimal if and only if*

1. *automaton A is minimal, that is, all states are reachable and there is no pair of different but equivalent states, and*

2. *there is no pair of different but almost-equivalent states, such that at least one of them is in the preamble.*

A simple method for constructing a hyper-minimal automaton from some given DFA is a state merging approach, upon which most hyper-minimization algorithms are based:

> as long as there are different but almost-equivalent states p and q, with $p \in \mathrm{Pre}(A)$, such that p is not reachable from q, merge p to q.

This algorithm always yields a hyper-minimal DFA that is almost-equivalent to the given one [5].

Example 3.2. The languages $L = aa^+ + b^*c$ and $L' = a^+ + b^*c$ are almost-equivalent, because their symmetric difference is the finite set $L \triangle L' = \{a\}$. Therefore, the two DFAs A and A', which are depicted in Figure 3.1, are almost-equivalent, because $L(A) = L$ and $L(A') = L'$. Further, since the right languages of the states q_2 and q_5 of DFA A are almost-equivalent, namely $L_A(q_2) = \{\lambda\}$ and $L_A(q_5) = \emptyset$, we have $q_2 \sim q_5$. The states q_1 and q_4 of A are not almost-equivalent, because their right languages are $L_A(q_1) = b^*c$ and $L_A(q_4) = a^*$, respectively, and the symmetric difference of these two languages is infinite.

The preamble of automaton A consists of the states q_0 and q_3, while the other states q_1, q_2, q_4, and q_5 are the kernel states of A. In the automaton A' we have $\mathrm{Pre}(A') = \{q_0'\}$ and $\mathrm{Ker}(A') = \{q_1', q_2', q_4', q_5'\}$.

Obviously A is not hyper-minimal, because $A \sim A'$ and A' has fewer states than A. We could also use Theorem 3.1 to see this: the states q_3 and q_4 are almost-equivalent, and q_3 is a preamble state, so Condition 2 of the theorem is not satisfied. We can also check that A' is hyper-minimal: since all states are reachable, and there are no two different but equivalent states, the DFA A is minimal. Further, the only preamble state q_0' is not almost-equivalent to any other state in A', so both conditions of Theorem 3.1 are satisfied.

In fact, the automaton A' is obtained from automaton A by merging state q_3 to state q_4—note that these two states are almost-equivalent, and state q_3 is a preamble state that is not reachable from q_4. $\diamond$

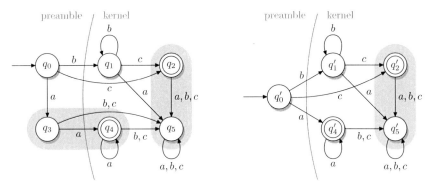

Figure 3.1: Two almost-equivalent minimal DFAs A (left) and A' (right), accepting the languages $L(A) = aa^+ + b^*c$ and $L(A') = a^+ + b^*c$. The gray shading denotes almost-equivalence of states.

Unlike in the classical case, where the minimal DFA for a regular language is unique, there may be several different almost-equivalent hyper-minimal DFAs for a given language—we will shortly see such an example in Figure 3.2 on page 33. Nevertheless, the structure of these automata is still very similar, as the following result from [5] shows. Roughly speaking, the kernels of two minimal almost-equivalent automata are isomorphic, and for hyper-minimal almost-equivalent automata, also the preambles are isomorphic, but the connections between preamble and kernel may be different.

Theorem 3.3. *Let $A = (Q, \Sigma, \delta, q_0, F)$ and $A' = (Q', \Sigma, \delta', q'_0, F')$ be two minimal deterministic finite automata with $A \sim A'$. Then there exists a mapping $h \colon Q \to Q'$ satisfying the following conditions.*

1. If $q \in \mathrm{Pre}(A)$, then $q \sim h(q)$, and if $q \in \mathrm{Ker}(A)$, then $q \equiv h(q)$.

2. If $q_0 \in \mathrm{Pre}(A)$, then $h(q_0) = q'_0$, and if $q_0 \in \mathrm{Ker}(A)$, then $h(q_0) \sim q'_0$.

3. The restriction of h to $\mathrm{Ker}(A)$ is a bijection between the kernels of A and A', that is compatible with taking transitions:

 3.a It is $h(\mathrm{Ker}(A)) = \mathrm{Ker}(A')$, and if $q_1, q_2 \in \mathrm{Ker}(A)$ with $h(q_1) = h(q_2)$, then $q_1 = q_2$.

 3.b It is $h(\delta(q, a)) = \delta'(h(q), a)$, for all $q \in \mathrm{Ker}(A)$ and all $a \in \Sigma$.

Moreover, if both automata A and A' are hyper-minimal, then also the following condition holds.

4. The restriction of h to $\mathrm{Pre}(A)$ is a bijection between the preambles of A and A', that is compatible with taking transitions, except for transitions from preamble to kernel states:

4.a It is $h(\mathrm{Pre}(A)) = \mathrm{Pre}(A')$, and if $q_1, q_2 \in \mathrm{Pre}(A)$ with $h(q_1) = h(q_2)$, then $q_1 = q_2$.

4.b It is $h(\delta(q, a)) = \delta'(h(q), a)$, for all $q \in \mathrm{Pre}(A)$ and all $a \in \Sigma$, that satisfy $\delta(q, a) \in \mathrm{Pre}(A)$.

Note that Condition 1 of Theorem 3.3 implies that for a kernel state q of A we have $q \in F$ if and only if $h(q) \in F'$, but this is not necessarily true for preamble states. Further, Conditions 3.b and 4.b say that transitions in the kernel and transitions in the preamble are preserved, but nothing is said about transitions leading from the preamble to the kernel. It could be that for some $q \in \mathrm{Pre}(A)$ and $a \in \Sigma$, with $\delta(q, a) \in \mathrm{Ker}(A)$, we have $h(\delta(q, a)) \neq \delta'(h(q), a)$. Thus, two almost-equivalent hyper-minimal DFAs can differ in the following aspects:

- acceptance values of preamble states,

- transitions leading from preamble to kernel states,

- the initial state, if the preambles of the automata are empty.

In case the preambles are empty, the initial states of the automata still have to be almost-equivalent. Further, also the transitions leading from preamble to kernel states cannot be changed arbitrarily. In fact, even if for some $q \in \mathrm{Pre}(A)$ and $a \in \Sigma$, with $\delta(q, a) \in \mathrm{Ker}(A)$, the two states $h(\delta(q, a))$ and $\delta'(h(q), a)$ are different, they still have to be almost-equivalent. This follows from the fact that $\sim$ is an equivalence relation, and the following result from [5].

Lemma 3.4. *Let* $A = (Q, \Sigma, \delta, q_0, F)$ *and* $A' = (Q', \Sigma, \delta', q_0', F')$ *be two (not necessarily distinct) deterministic finite automata, with* $q \in Q$ *and* $q' \in Q'$. *Then* $q \sim q'$ *if and only if* $\delta(q, w) \sim \delta'(q', w)$, *for all* $w \in \Sigma^*$.

Example 3.5. Consider again the two DFAs from Figure 3.1 on the previous page, and the mapping h defined as follows:

$$h(q_0) = q_0', \quad h(q_1) = q_1', \quad h(q_2) = q_2', \quad h(q_3) = h(q_4) = q_4', \quad h(q_5) = q_5'.$$

This mapping satisfies Conditions 1, 2, and 3 of Theorem 3.3, in particular, the kernels of both automata are isomorphic. Since the preambles of the two automata are not isomorphic, Condition 4 is not satisfied.

Now compare the DFAs A' and A'', which are depicted in Figure 3.2. Automaton A'' accepts the language $L(A'') = a^* + b^+ c$, and is almost-equivalent to A' (thus, also to A). Further, both A' and A'' are hyper-minimal. Now the obvious mapping g, with $g(q_i') = q_i''$, satisfies all conditions of Theorem 3.3. The automata A' and A'' only differ in the acceptance of their single preamble states q_0' and q_0'', and in the transitions on input c from these preamble states into the kernels of the automata: we have $g(\delta'(q_0', c)) = g(q_5') = q_5''$, but $\delta''(g(q_0'), c) = \delta''(q_0'', c) = q_2''$—here δ' and δ'' are the transition functions of A' and A'', respectively. Nevertheless, the states q_2'' and q_5'' are almost-equivalent. $\diamond$

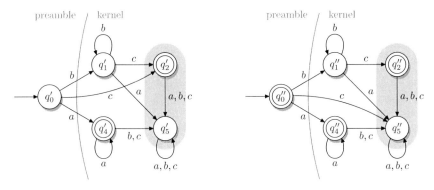

Figure 3.2: The two almost-equivalent hyper-minimal DFAs A' (left) and A'' (right). The gray shading denotes almost-equivalence of states.

On the Error Length Between Almost-Equivalent Languages

Since the symmetric difference between any two almost-equivalent languages L and L' is a *finite* set, the length of the words in $L \triangle L'$ is bounded. The question how this bound depends on the languages L and L' naturally occurs. In the case of regular languages, an upper bound for this length can be derived from the sizes of the DFAs for L and L'. In papers of Gawrychowski and Jeż [34], and Gawrychowski et al. [35], the *distance* between two languages L and L' is defined as follows:

$$d(L, L') = \begin{cases} 0 & \text{if } L = L', \\ 1 + \max\{ |w| \mid w \in L \triangle L' \} & \text{if } L \neq L' \text{ and } L \sim L', \\ \infty & \text{if } L \nsim L'. \end{cases}$$

This distance function satisfies the following *strong triangle inequality*—in fact, function d is a so-called *ultrametric*.

Lemma 3.6. *For all languages L, L', and L'' we have*

$$d(L, L'') \leq \max(d(L, L'), d(L', L'')).$$

Proof. Clearly, for $L = L''$ the statement holds. Next assume $L \nsim L''$, which means that there are infinitely many words in $L \setminus L''$ or in $L'' \setminus L$. For symmetry reasons, we only need to consider the case where $|L \setminus L''| = \infty$. Either infinitely many words from $L \setminus L''$ do not belong to L', or all but a finite number of words from $L \setminus L''$ belong to L'. In the first case we obtain $d(L, L') = \infty$ because $L \nsim L'$, and in the second case we obtain $d(L', L'') = \infty$. Hence $d(L, L'') \leq \max(d(L, L'), d(L', L'')) = \infty$.

Finally assume $L \sim L''$ but $L \neq L''$. Consider a longest word $w \in L \triangle L''$. Either $w \in L \setminus L''$ or $w \in L'' \setminus L$. For the case $w \in L \setminus L''$ we consider the following two sub-cases: if $w \in L'$, then $|w| \leq d(L', L'') - 1$, and if $w \notin L'$, then $|w| \leq d(L, L') - 1$. Hence, we obtain $|w| \leq \max(d(L, L'), d(L', L'')) - 1$. A similar argument for the case $w \in L'' \setminus L$ gives the same inequality. Therefore $d(L, L'') \leq \max(d(L, L'), d(L', L''))$. $\qquad\square$

The *distance* between two states q and q' of finite automata A and A' is then defined by $d(q, q') = d(L_A(q), L_{A'}(q'))$. Obviously, two languages L and L' (or states q and q') are almost-equivalent if and only if $d(L, L') < \infty$ (or, respectively, $d(q, q') < \infty$). The following result can be found in the papers of Gawrychowski and Jeż [34] and Gawrychowski et al. [35].

Lemma 3.7. *Let $A = (Q, \Sigma, \delta, q_0, F)$ be a deterministic finite automaton, and let $q, q' \in Q$. Then either $d(q, q') = \infty$ or $d(q, q') \leq |Q| - 1$.*

Note that above result only considers states in a single automaton. Given two DFAs A and A' having n and n' states, respectively, one could think of them as *one* DFA with $n + n'$ states, by simply placing one beside the other. Let q_0 and q_0' be the respective initial states of the DFAs A and A'. Now Lemma 3.7 implies that $d(q_0, q_0') < n + n'$, hence the length of the longest words in $L(A) \triangle L(A')$ is at most $n + n' - 2$. In fact, one can reduce this bound to $m - 2$, where $m = \max(n, n')$, as the following result shows.

Theorem 3.8. *Let L and L' be regular languages accepted by deterministic finite automata having n, and n' states, respectively, and let $m = \max(n, n')$. If $L \sim L'$, then $d(L, L') \leq m - 1$, that is, $\max\{|w| \mid w \in L \triangle L'\} \leq m - 2$.*

Proof. Let $A = (Q, \Sigma, \delta, q_0, F)$ and $A' = (Q', \Sigma, \delta', q_0', F')$ be two almost-equivalent DFAs with $|Q| = n$ and $|Q'| = n'$, and let $L = L(A)$ and $L' = L(A')$.

We first investigate the case, where the automata A and A' are hyperminimal. Let $h \colon Q \to Q'$ be the mapping from Theorem 3.3, and assume that $w \in L \triangle L'$. If the preambles of both automata are empty, then A and A' are isomorphic up to the choice of the initial state. We can use Lemma 3.7 on the states $h(q_0)$ and q_0' of the automaton A' and obtain $d(h(q_0), q_0') \leq n' - 1 = m - 1$. Since $L' = L_{A'}(q_0')$ and $L = L_{A'}(h(q_0))$—recall that $h(q_0) \equiv q_0$—the statement follows. If the preambles of A and A' are not empty and w leads to a preamble state in both automata—note that by Theorem 3.3 the word w leads to a preamble state in A if and only if it leads to a preamble state in A'—then the length of w is bounded by $n - 2$, and the statement follows. Next, we consider words leading to kernel states of A and A'. Let $w = uav$, for $u, v \in \Sigma^*$ and $a \in \Sigma$, such that the state $q_u = \delta(q_0, u)$ is still a preamble state and $q_{ua} = \delta(q_0, ua)$ is the first kernel state in the computation of A on input w. By Theorem 3.3, also the state $q_u' = \delta'(q_0', u)$ is a preamble state, and the state $q_{ua}' = \delta'(q_0', ua)$ is in the kernel of A'. Note that $w = uav \in L \triangle L'$ implies $v \in L(q_{ua}) \triangle L(q_{ua}')$. Now consider the two DFAs B and B' obtained from A and A' by choosing the states q_{ua} and q_{ua}', respectively, as initial states. The preambles of B and B' can be considered empty, and the number of states of B (and of B') is at most $|\mathrm{Ker}(A)|$. Since B and B' are isomorphic up to the choice of initial states, we can use Lemma 3.7 again to obtain $|v| \leq |\text{Ker}(A)| - 2$. Further, since q_u is a preamble state, we have $|u| \leq |\text{Pre}(A)| - 1$. Hence $|w| = |uav| \leq n - 2$, which proves the statement.

Now assume that A is not hyper-minimal, and A' results from A by merging state $q \in \text{Pre}(A)$ to some state q', where $q \sim q'$, and q is not reachable from q'. If $w \in L \triangle L'$, then the computation in A on input w must pass through state q, because otherwise the same computation could be carried out in A', which would contradict $w \in L \triangle L'$. Hence w can be written as $w = w_1 w_2$, for $w_1, w_2 \in \Sigma^*$, such that $\delta(q_0, w_1) = q$ and $\delta'(q'_0, w_1) = q'$, and further $w_2 \in L(q) \triangle L(q')$. Let $P \subseteq Q$ be the set of states that A passes through by reading a strict prefix of w_1. Consider the automaton B that is obtained from A by choosing initial state q. Since q is a preamble state, none of the states from P are reachable in B. Further, because q is not reachable from q', also no state from P is reachable from state q'. Hence the states from P can safely be removed from automaton B. Now we have two almost-equivalent states q and q' in an automaton with at most $|Q| - |P|$ states. Applying Lemma 3.7 on the word $w_2 \in L(q) \triangle L(q')$ gives $|w_2| \leq |Q| - |P| - 2 \leq n - |w_1| - 2$. It follows $|w| \leq n - 2 = \max(n, n') - 2$ because $n' = n - 1$.

We have shown that the statement of the theorem holds if both languages L and L' are accepted by almost-equivalent hyper-minimal DFAs, and also if L' results from L by performing a single state merging step. To put everything together, assume that A and A' are arbitrary almost-equivalent DFAs. By applying a state merging algorithm to both automata, we obtain two almost-equivalent hyper-minimal DFAs A_* and A'_* accepting the languages L_* and L'_*, respectively. From the first part of the proof we know $d(L_*, L'_*) \leq m - 1$. By iteratively using the argumentation from the second part of the proof we get $d(L, L_*), d(L'_*, L') \leq m - 1$. By Lemma 3.6 now follows $d(L, L') \leq m - 1$, which concludes our proof. $\qquad\square$

The bound on the length of the longest word in $L \triangle L'$ for almost-equivalent regular languages L and L' is easily seen tight by the witness languages $L = \emptyset$ and $L' = \{a^{n-2}\}$.

3.2 E-Equivalence and E-Minimality

Almost-equivalence allows for a finite number of errors to be made, but there is no control about which errors may or may not occur. On the other hand, classical equivalence allows no errors at all. We generalize the notion of equivalence and almost-equivalence by introducing a concept that explicitly parameterizes the difference, or error language, that is allowed between related languages. In this way many known equivalence notions from the literature can be covered, as we will see in Section 3.3.

Let E be any subset of Σ^*, called the *error language*. Then we say that two languages L_1 and L_2 over the alphabet Σ are equivalent with respect to the error language E, for short E-*equivalent*, if their symmetric difference lies in E, that is, if $L_1 \triangle L_2 \subseteq E$. In this case we write $L_1 \sim_E L_2$. As for equivalence and almost-equivalence, the E-equivalence relation naturally carries over to finite automata and states. To finite automata A and A' are E-*equivalent* if $L(A) \sim_E L(A')$, and two states q and q' of automata A and A', respectively, are E-*equivalent* if $L_A(q) \sim_E L_{A'}(q')$. We write $A \sim_E A'$ and $q \sim_E q'$ to denote E-equivalence between automata A and A', and, respectively, states q and q'.

The following result gives a characterization of E-equivalent languages in terms of classical language equality.

Lemma 3.9. *We have* $L \sim_E L'$ *if and only if* $L \cup E = L' \cup E$.

Proof. By definition, the relation $L \sim_E L'$ holds if and only if $L \triangle L' \subseteq E$, which is equivalent to $(L \triangle L') \setminus E = \emptyset$. Since both sets $L \cup E$ and $L' \cup E$ contain all words from E, they can only differ in words from $(L \triangle L') \setminus E$, that means, we have $(L \cup E) \triangle (L' \cup E) = (L \triangle L') \setminus E$. Thus, $L \sim_E L'$ holds if and only if $L \cup E = L' \cup E$. □

Since equality of languages is an equivalence relation, the following is a direct consequence of Lemma 3.9.

Corollary 3.10. *The relation* $\sim_E$ *is an equivalence relation.* □

A deterministic (or nondeterministic) finite automaton A is E-*minimal* if there does not exist a deterministic (or nondeterministic, respectively) finite automaton A' with fewer states than A, that satisfies $A \sim_E A'$.

Unfortunately, there is no structural similarity between different E-equivalent finite automata, as it is the case for equivalence or almost-equivalence. This will illustrated in the upcoming example. Note that any two languages can be set in relation to each other by choosing an appropriate error set, that is, any two DFAs can be E-equivalent. Of course, the larger the set of allowed errors is, the more languages are E-equivalent to each other. Therefore, the chance of finding even smaller E-equivalent automata should grow with the size of E. But this does *not* mean that an E-minimal automaton which is E-equivalent to some given automaton makes use of as much errors as possible.

Example 3.11. Consider the two languages $L_1 = (a^7)^*$ and $L_2 = a^2 + (a^3)^*$. If we want these languages to be E-equivalent, then the error set E as to satisfy

$$
\begin{aligned}
E &\supseteq L_1 \triangle L_2 \\
&= [(a^7)^* \setminus (a^2 + (a^3)^*)] \cup [(a^2 + (a^3)^*) \setminus (a^7)^*] \\
&= [(a^7)^* \setminus (a^3)^*] \cup [(a^3)^* \setminus (a^7)^*] \cup a^2.
\end{aligned}
$$

Thus, by the choice of an appropriate error language E one observes that also the language $L_3 = (a^3)^*$ is E-equivalent to both languages L_1 and L_2. Hence, if $L_1 \sim_E L_2$, for some error language E, then also $L_1 \sim_E L_3$ and $L_2 \sim_E L_3$. The minimal DFAs accepting the languages L_1, L_2, and L_3 are depicted in Figure 3.3. Observe that the minimal DFA accepting the language L_1 has

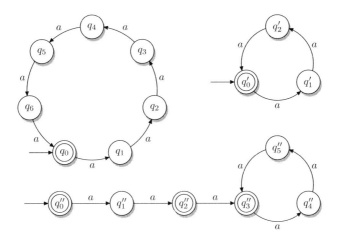

Figure 3.3: Minimal DFAs (from left to right and top to bottom) accepting the languages $L_1 = (a^7)^*$, $L_3 = (a^3)^*$, and $L_2 = a^2 + (a^3)^*$, which are E-equivalent for every error set E that satisfies $E \supseteq L_1 \triangle L_2$.

seven states, while that for accepting L_2 has six states, and the automaton for language L_3 has three states only. $\Diamond$

Since there is no structural similarity between E-equivalent automata, there is also no structural characterization of E-minimal DFAs. In the case of equivalence and almost-equivalence, such characterizations help to find efficient minimization algorithms. For the problem of E-minimization of finite automata, we do not hope to find an efficient algorithm. In fact, we will see in Chapter 4 that the problem of E-minimization is computationally intractable, even for DFAs, and even if the given error language is finite.

Before we look at some more error profiles in the next section, we shortly discuss a connection between almost-equivalence and E-equivalence. By definition, two languages L and L' over an alphabet Σ are almost-equivalent if their symmetric difference is finite. This means that there exists *some* finite error set $E \subseteq \Sigma^*$, such that L and L' are E-equivalent. In fact, in the case of regular languages an appropriate set E can be derived as follows: Let n and n' be the numbers of states of the (minimal) DFAs for L and L', and $m = \max(n, n')$. By Theorem 3.8 we know that $L \sim L'$ implies that the length of the longest words in $L \triangle L'$ is at most $m - 2$. Therefore we obtain the following result.

Theorem 3.12. *Let L and L' be regular languages over alphabet Σ, which are accepted by deterministic finite automata having n and n' states, respectively. Further, let $m = \max(n, n')$ and $E = \Sigma^{\leq m-2}$. Then $L \sim L'$ holds if and only if $L \sim_E L'$. Moreover, if q and q' are states of a deterministic finite automaton with n states, then $q \sim q'$ if and only if $q \sim_E q'$, for $E = \Sigma^{\leq n-2}$.* $\square$

Notice that almost-equivalence of the languages L and L' from Theorem 3.12 also implies E-equivalence for all error sets $E = \Sigma^{\leq \ell}$ with $\ell \geq m - 2$. In a similar way, almost-equivalence of states q and q' from Theorem 3.12 implies E-equivalence for all sets $E = \Sigma^{\leq \ell}$ with $\ell \geq n - 2$.

3.3 Further Error Profiles

This section presents further error profiles, which are somehow related to almost-equivalence, and which can also be covered by the concept of E-equivalence.

We start with the notion of k-equivalence, which was already considered by Badr et al. [5] as a direction for further research, and for which efficient minimization algorithms were presented by Gawrychowski et al. [34, 35]. In the mentioned literature, k-equivalence is also called k-f-equivalence, or k-similarity. Two languages L and L' over an alphabet Σ are k-equivalent, denoted by $L \sim_k L'$, if they differ only in words of length at most k, that is, if $L \triangle L' \subseteq \Sigma^{\leq k}$. Two finite automata A and A' are k-equivalent, denoted by $A \sim_k A'$, if $L(A) \sim_k L(A')$. One could of course also define k-equivalence for states of finite automata, but for the known minimization algorithms [34, 35] a different relation, namely that of k-similarity, is used. We will describe this relation in more detail in Section 5.5, where a k-minimization algorithm for DFAs, which is based on Brzozowski's DFA minimization algorithm, will be presented. Finally, a deterministic (or nondeterministic) finite automaton A is k-minimal if there is no deterministic (or nondeterministic, respectively) finite automaton B, that has fewer states than A, and satisfies $A \sim_k B$.

Obviously the notions of k-equivalence and k-minimality can be expressed by E-equivalence and E-minimality, when choosing $E = \Sigma^{\leq k}$. Notice also, that k-equivalence of languages or automata implies almost-equivalence, since $\Sigma^{\leq k}$ is a finite set.

Example 3.13. Consider the deterministic finite automata A and A' which are depicted in Figure 3.4 and, respectively, Figure 3.5. The symmetric difference between the two languages $L(A)$ and $L(A')$ is

$$L(A) \triangle L(A') = (b + aa + ab)(b + aa + ab)(\lambda + a),$$

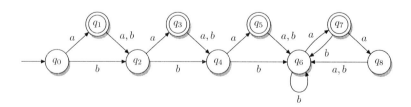

Figure 3.4: The deterministic finite automaton A for Example 3.13.

which can be seen as follows: words from $(b+aa+ab)(b+aa+ab)$ take automaton A to state q_4, and automaton A' to state q_7', and the symmetric difference between the right languages of these states is $L_A(q_4) \triangle L_{A'}(q_7') = \{\lambda, a\}$. In particular, the longest words in $L(A) \triangle L(A')$ have length 5. Therefore, the two languages are k-equivalent, for $k = 5$, and so $A \sim_5 A'$. This can also be expressed as E-equivalence, namely $A \sim_E A'$, for the error language $E = \Sigma^{\leq 5}$. In fact, the automaton A' is also k-minimal, because it is already hyper-minimal—this follows from Theorem 3.1, and the fact that no two states in A' are almost-equivalent. $\diamondsuit$

A different error profile, which is complementary to k-equivalence, is the concept of cover languages and cover automata. Deterministic cover automata were introduced by Cezar Câmpeanu, Nicolae Sântean, and Sheng Yu [17], first results on nondeterministic cover automata were recently presented by Câmpeanu [16]. Cover automata are used to succinctly represent finite languages by allowing the automaton to accept, besides the words from the original language, also words that are longer than any word in the given language. If $L \subseteq \Sigma^*$ is a finite language, and ℓ is the length of the longest word in L—this number is called the *cover length*—then a language $L' \subseteq \Sigma^*$ is called a *cover language* for L if $L' \cap \Sigma^{\leq \ell} = L$. Further, a finite automaton A is a *cover automaton* for L if $L(A)$ is a cover language for L, that is, if $L(A) \cap \Sigma^{\leq \ell} = L$. The automaton A is a *minimal cover automaton*, or *cover-minimal* for L, if A is a cover automaton for L and there is no other cover automaton B (of the same type as A) for L, that has fewer states than A. Efficient algorithms for constructing minimal deterministic cover automata are given by Cezar Câmpeanu, Andrei Păun, and Jason R. Smith [19], Câmpeanu et al. [18], and Heiko Körner [76]. Another algorithm for this problem, which is similar to Brzozowski's DFA minimization algorithm, will be described in Section 5.4.

Also the notion of cover languages and automata can be seen as a special case of E-equivalence. Let $L \subseteq \Sigma^*$ be a finite language, and ℓ be the cover length of L. Every language L' that is a cover language for L satisfies $L' \cap \Sigma^{\leq \ell} = L$. Note that also L itself is a cover language, since $L \cap \Sigma^{\leq \ell} = L$ by definition of ℓ. Hence, the condition $L' \cap \Sigma^{\leq \ell} = L$ can also be written as $L' \cup \Sigma^{> \ell} = L \cup \Sigma^{> \ell}$. Now it can readily be seen that two languages L' and L'' over an alphabet Σ

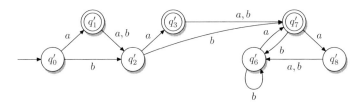

Figure 3.5: The deterministic finite automaton A' that is k-equivalent, for $k = 5$, to the finite automaton A from Figure 3.4.

are cover languages for a finite language L with cover length ℓ, if and only if $L' \sim_E L''$, for $E = \Sigma^{>\ell}$—recall that $L' \sim_E L''$ if and only if $L' \cup E = L'' \cup E$ by Lemma 3.9. This means that an automaton A is cover-minimal if and only if it is $\Sigma^{>\ell}$-minimal.

Example 3.14. Consider again the automata A and A' from Example 3.13. Recall that these automata are E-equivalent, for $E = \Sigma^{\leq 5}$, which means that automaton A' correctly accepts all words from $L(A)$ that are longer than 5 symbols. If we want to accept shorter words from $L(A)$ correctly with another automaton, we could use a cover automaton for the language $L(A) \cap \Sigma^{\leq 5}$. The DFA A'', depicted in Figure 3.6, is a cover automaton for that language, because the shortest word in $L(A) \triangle L(A'')$ is of length 6. One can easily see that no single-state DFA can be a cover automaton for $L(A) \cap \Sigma^{\leq 5}$, hence automaton A'' is a cover-minimal automaton for $L(A)$. $\diamond$

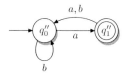

Figure 3.6: The cover-minimal automaton A'' for the language $L(A) \cap \Sigma^{\leq 5}$.

Besides almost-equivalence, k-equivalence, and cover languages there may be further concepts in the literature which can be described with the notion of E-equivalence. For example, sometimes languages are regarded equivalent, if they differ at most in the empty word λ—here, we have $E = \{\lambda\}$. Another concept is that of closeness. Two languages are *close* if they are E-equivalent, for some *sparse* language E—a language E is sparse, if there is a polynomial p such that $|E \cap \Sigma^n| \leq p(n)$, for every $n \geq 0$.

In the following chapter we will consider problems concerning the error profiles of almost-equivalence and E-equivalence, and Chapter 5 will also present results on cover automata and k-equivalent automata.

Chapter 4

Computational Complexity of Problems Related to Minimal Finite Automata

THE study of the computational complexity of problems on descriptional models for regular languages dates back to works of Meyer and Stockmeyer from the early 1970s [95, 112]. A survey of recent results on the computational complexity of problems concerning finite automata can be found in a paper of Holzer and Kutrib [58]. In this chapter we investigate the computational complexity of problems for finite automata, which are related to equivalence and minimality with respect to the error profiles of almost-equivalence and E-equivalence. While there will be no difference in the complexity of equivalence problems for the different error profiles, which will be studied in Section 4.1, we will see in Section 4.2 that the complexity of minimization problems depends on the chosen error profile. Also the results on canonicity problems, which will be discussed in Section 4.3, show a difference between complexities of the different error profiles. All problems in the aforementioned sections are decision problems. In contrast, in Section 4.4 we consider counting problems for minimal finite automata: since the studied error profiles allow several different minimal DFAs for a given regular language, we want to *count* the number of these minimal deterministic automata.

Before we start our investigations on finite automata problems for different error profiles, we recall two decision problems for finite automata, which will be referred to throughout this chapter. The *universality problem* for nondeterministic finite automata is defined as follows:

given a nondeterministic finite automaton A over the alphabet Σ, decide whether $L(A) = \Sigma^*$ holds.

This problem is know to be PSPACE-complete, which follows from corresponding results on regular expressions as presented by Aho et al. [1], and Meyer and

Stockmeyer [95]. A similar problem is the *union universality problem* for DFAs:

given some deterministic finite automata $A_1, A_2, \ldots, A_k$ over a common alphabet Σ, decide whether $\bigcup_{i=1}^{k} L(A_i) = \Sigma^*$.

This problem can be seen to be PSPACE-complete by a reduction from the PSPACE-complete intersection emptiness problem for deterministic finite automata, which asks whether the intersection of the given automata is non-empty—see the papers of Jiang and Ravikumar [67], and Dexter Kozen [77].

4.1 Equivalence Problems

The most basic problem for automata is that of ordinary equivalence. This is the problem of deciding for two given automata A and B, whether $A \equiv B$ holds. The complexity of this classical problem is well known. For DFAs the problem is NL-complete—see for example the paper of Cho and Huynh [22]—and for NFAs it is PSPACE-complete, which follows from results of Aho et al. [1], and Meyer and Stockmeyer [95]. We show that this situation is resembled for the almost-equivalence and the E-equivalence problem.

Theorem 4.1 (Almost-Equivalence Problem). *The problem of deciding for two given finite automata A and B, whether $A \sim B$, is* PSPACE-*complete if at least one of the automata A and B is nondeterministic, and it is* NL-*complete if both automata A and B are deterministic.*

Proof. We first consider the case, where A and B are deterministic. Containment in NL is seen as follows: on input A and B we construct a DFA for the symmetric difference of $L(A)$ and $L(B)$ by standard methods. Then we check whether this automaton accepts a finite language. The automaton accepting $L(A) \triangle L(B)$ can be constructed in deterministic logspace. Since it was shown by Jones [69] that the problem of deciding whether a given automaton accepts a finite language is NL-complete, we obtain the claimed upper bound. From this problem we also get the lower bound by the following reduction: let A be some given DFA and let B be the single-state DFA accepting the empty language $\emptyset$. Then $A \sim B$ if and only if A accepts a finite language, which proves NL-hardness.

Now we turn to the case where at least one automaton is nondeterministic. We will describe how containment of the problem in PSPACE can be concluded from the NL upper bound of the corresponding problem for DFAs. This technique will also be used in several other proofs, and will then be simply referred to as *on-the-fly power-set construction*.

Let A and B be two nondeterministic finite automata for which we want to decide, whether $A \sim B$ holds. Further, let $A' = \mathcal{P}(A)$ and $B' = \mathcal{P}(B)$ be the corresponding power-set DFAs of A and B, respectively. Of course $A \sim B$ if and only if $A' \sim B'$. So instead of deciding whether A and B are almost-equivalent, we may also decide this question for their power-set automata A'

and B'. This can be done by a nondeterministic Turing machine with a space bound that is logarithmic in the size of its input A' and B'. To obtain a PSPACE algorithm for solving the almost-equivalence problem for NFAs we now combine the power-set construction with the NL algorithm for the corresponding DFA problem. Whenever the NL algorithm wants to read a symbol from its input, that is, from the descriptions of A' and B', we generate this part of the input on a special working tape of our PSPACE Turing machine, by running the power-set construction on the input NFAs. Instead writing down the whole power-set automaton on the working tape, which would need exponential space, we only write the part which the NL algorithm wants to read from its input. To keep track of the position in the input the NL algorithm wants to read, we use a binary counter, which can be implemented using logarithmic space in the size of the input for the NL algorithm. Also the space used by the NL algorithm on its own working tape is logarithmic in the size of its input. Since the input to the NL algorithm are the representations of the power-set DFAs A' and B', this size is at most exponential in the size of the description of the NFAs A and B. Therefore the space used by the combined algorithm is at most polynomial in the size of the descriptions of the input NFAs A and B. Hence, we the problem can be solved in PSPACE.

To show PSPACE-hardness, we give a reduction from the universality problem for NFAs. Let the NFA $C = (Q, \Sigma, \delta, q_0, F)$ be an instance of this problem. We define the NFA $A = (Q, \Sigma \cup \{\#\}, \delta', q_0, F)$, where $\# \notin \Sigma$ is a new input symbol, and the transition function δ' is defined such that $\delta'(q, a) = \delta(q, a)$, for every state $q \in Q$ and symbol $a \in \Sigma$, and $\delta'(q, \#) = \{q_0\}$, for every state $q \in Q$. Further, let B be the single-state DFA accepting the language $(\Sigma \cup \{\#\})^*$. Clearly, this construction can be carried out by a logarithmic space bounded Turing machine.

It is easy to see that if $L(C) = \Sigma^*$, then the constructed NFA A accepts the language $(\Sigma \cup \{\#\})^*$, and thus is almost-equivalent to B. Otherwise, if $L(C) \neq \Sigma^*$, there is a word $w \in \Sigma^* \setminus L(C)$. Then one readily verifies that by construction of A, the infinite set of words $w(\#w)^*$ is not included in the language $L(A)$. Therefore, there are infinitely many words in $L(B) \setminus L(A)$. This shows that $L(C) = \Sigma^*$ holds if and only if $A \sim B$, which gives the desired PSPACE lower bound, and concludes our proof. □

We now turn to the problem of deciding E-equivalence of finite automata. For this problem, and all further problems dealing with E-equivalence, a finite automaton A_E that specifies the error language E is given as additional input.

Theorem 4.2 (E-Equivalence Problem). *The problem of deciding for given finite automata A, B, and A_E, whether $A \sim_E B$, for $E = L(A_E)$, is PSPACE-complete if at least one of the automata A, B, and A_E is nondeterministic, and it is NL-complete if all three automata A, B and A_E are deterministic.*

Proof. Again, we start with the case where all given automata are deterministic. The NL lower bound immediately follows from the lower bound on ordinary equivalence [22], because $\equiv$ is a special case of E-equivalence when choosing the error language $E = \emptyset$.

The upper bound is seen as follows: let finite automata A, B, and A_E be given, and let $E = L(A_E)$. We construct two automata A' and B' accepting the languages $L(A) \cup E$ and $L(B) \cup E$, respectively, by a cross product construction. This construction can be performed by a logarithmic space bounded Turing machine. Since by Lemma 3.9 $L(A) \sim_E L(B)$ if and only if $L(A) \cup E = L(B) \cup E$, we have $A \sim_E B$ if and only if $A' \equiv B'$. This gives a logspace reduction from the E-equivalence problem to the ordinary equivalence problem.

With a similar reasoning we can derive a PSPACE upper bound for the problem, where at least one of the involved automata is an NFA. From the given automata A, B, and A_E we can construct two NFAs A' and B' accepting the languages $L(A) \cup E$ and $L(B) \cup E$, using only polynomial space. By Lemma 3.9, we know that $A \sim_E B$ if and only if $A' \equiv B'$. The latter equivalence can be decided in PSPACE [95].

It remains to prove PSPACE-hardness for the case where at least one automaton is nondeterministic. Here we give a reduction from the universality problem for NFAs. As an instance of this problem let A_1 be an NFA over the alphabet Σ. Further, let A_2 be the single-state DFA for the language Σ^*, and A_3 be the single-state DFA for the empty language $\emptyset$. Then A_1 is a positive instance of the universality problem if and only if $A_2 \sim_E A_3$, for $E = L(A_1)$, which shows PSPACE-hardness of the E-equivalence problem if the error language E may be specified by an NFA. Now assume the error language must be specified by a DFA, and one of the other two input automata for the E-equivalence problem may be an NFA. Then PSPACE-hardness can be seen by the fact that A_1 is a positive instance of the universality problem if and only if $A_1 \sim_E A_2$, for $E = L(A_3) = \emptyset$. Obviously, these reductions can be carried out by logarithmic space bounded Turing machines. $\square$

We summarize the complexities of the different equivalence problems in Table 4.1. Note that the problem of deciding, whether two deterministic finite automata are E-equivalent, is already PSPACE-complete if only the automaton A_E that specifies the error language is an NFA.

Equivalence Problem		
Equivalence relation	DFA	NFA
$\equiv$		
$\sim$	NL	PSPACE
$\sim_E$, for DFA A_E with $E = L(A_E)$		
$\sim_E$, for NFA A_E with $E = L(A_E)$	PSPACE	

Table 4.1: Results on the computational complexity of the equivalence, almost-equivalence, and E-equivalence problem. An entry indicates that the corresponding problem is complete for the stated complexity class.

The obtained complexity results naturally carry over to the corresponding problems of testing whether two states in an automaton are equivalent with respect to different error profiles. For example, it is NL-complete to decide for a given DFA A and two of its states p and q, whether $p \sim q$ holds.

4.2 Minimization and Minimality Problems

A classical problem in the theory of finite automata is the problem of finding an automaton that is as small as possible, but still equivalent to a given one. The decision version of the ordinary *minimization problem for finite automata* is the following:

> given a finite automaton A and an integer n, is there an n-state finite automaton B, with $A \equiv B$?

Here the automata A and be B may be deterministic or nondeterministic. For the NFA-to-DFA minimization, which is the problem where A is an NFA and B is a DFA, the integer n is given in unary notation. This is because the minimal DFA for a language can be exponentially larger than a given NFA for that language—see for example the papers of Meyer and Fischer [94], and Moore [97]. For the other cases the integer n may as well be given in binary notation. The DFA-to-DFA minimization problem, which is the problem where both automata A and B are deterministic, is computationally *easy*, namely NL-complete [22].

Since nondeterministic finite automata allow exponentially more succinct representations of regular languages than deterministic automata, also the minimization of nondeterministic machines is of practical interest. Unfortunately, descriptional succinctness often goes hand in hand with increased computational complexity. In fact the minimization problem becomes PSPACE-complete if nondeterministic finite automata are involved [67], even for very restricted types of nondeterminism, as studied by Henrik Björklund and Wim Martens [9], and Andreas Malcher [86]. This situation is resembled for the *minimality problem for finite automata*, which asks whether some given finite automaton is a minimal automaton. Also this problem is NL-complete for deterministic finite automata [22], and PSPACE-complete for nondeterministic machines [67].

In this section we study the complexity of the corresponding minimization and minimality problems for the error profiles of almost-equivalence and E-equivalence. The results on DFA problems are presented in the next subsection, and the NFA problems are discussed in Subsection 4.2.2.

4.2.1 Minimal Deterministic Finite Automata

In the following we show that the complexity of the hyper-minimization and hyper-minimality problems for DFAs is the same as the complexity for the corresponding classical problems: all these problems turn out to be NL-complete. However, when considering E-minimal automata, the complexity increases to

NP-completeness for the E-minimization problem, and coNP-completeness for the E-minimality problem.

We first consider the DFA-to-DFA hyper-minimization problem. For proving the upper bound, we have to check equivalence and almost-equivalence of states in deterministic finite automata, as well as to decide whether a given state is a preamble or a kernel state. The checks for equivalence and almost-equivalence can be done in NL, as described at the end of Section 4.1.

Deciding for a given DFA A and a state q of A, whether it belongs to the kernel or the preamble of the automaton can also be solved in NL as follows. Recall that q is a kernel state if and only if it is reachable from the initial state q_0 by an infinite number of words. This means that in the state transition graph of A there must be a path from the initial state q_0 to some state p, a (non-empty) path from state p to itself, and a path from state p to state q. After nondeterministically guessing an appropriate looping state p, the before mentioned reachability conditions can be checked in NL, which gives an NL algorithm for deciding whether a given state q is a kernel state. Since q is a preamble state if and only if it is *not* a preamble state, by the NL = coNL result of Immerman [65] and Szelepcsényi [113] also the question whether a state is a preamble state can be decided in NL.

Theorem 4.3 (Hyper-Minimization Problem). *The problem of deciding for a given deterministic finite automaton A and an integer n, whether there exists a deterministic finite automaton B with n states, such that $A \sim B$, is NL-complete.*

Proof. For containment in NL, we describe an algorithm that counts the number of states that are needed by any hyper-minimal automaton B, with $A \sim B$. If this number is at most n, then the answer is "yes," otherwise the answer is "no." Let $q_1, q_2, \ldots q_m$ be some fixed order of the states in A. To count the number of kernel states an almost-equivalent hyper-minimal DFA B needs, we iterate the following algorithm over all states q_i in ascending order. If $q_i \in \mathrm{Ker}(A)$, check whether $q_i \not\equiv q_j$ holds for all $j < i$. If this is the case, we increase the counter. This algorithm essentially counts the number of kernel states in the *minimal* DFA for $L(A)$, but since all hyper-minimal almost-equivalent DFAs have isomorphic kernels—cf. Theorem 3.3—this number is the same for all hyper-minimal DFAs that are almost-equivalent to A. To count the number of preamble states an almost-equivalent hyper-minimal DFA B needs, we iterate the following algorithm over all states q_i in ascending order. If $q_i \in \mathrm{Pre}(A)$, check whether $q_i \not\sim q_j$ holds for all $j < i$. If this is the case, then check whether $q_i \not\sim q_k$ for all kernel states q_k with $k > i$. If this is also the case, then increase the counter. This algorithm counts the number of almost-equivalence classes that consist of preamble states only. The reader is invited to verify that the described algorithm can be carried out on logarithmic space.

For proving NL-hardness we present a reduction from the classical minimization problem for DFAs. As an instance of that problem let A be a DFA over an alphabet Σ, and n be an integer. We construct a DFA A' from A

by adding a new input symbol $\# \notin \Sigma$, on which every state goes back to the initial state. The language accepted by A' is $(\Sigma^* \cdot \{\#\})^* L(A)$. If there exists an n-state DFA B, with $A \equiv B$, then we can construct from it a DFA B' by adding transitions on the new input $\#$ from every state back to the initial state. Since $L(B') = (\Sigma^* \cdot \{\#\})^* L(B) = L(A')$, there does exist an n-state DFA B' with $A' \sim B'$. For the converse implication assume there exists an n-state DFA B', with $A' \sim B'$. Then, since all states in A' are kernel states, we know by Theorem 3.3 that B' must be isomorphic (up to the choice of the initial state) to an n-state DFA B'', with $B'' \equiv A'$, so there is an n-state DFA B'' accepting the language $L(B'') = (\Sigma^* \cdot \{\#\})^* L(A)$. By removing all $\#$-transitions in B'' we obtain an n-state DFA B, satisfying $L(B) = L(A)$. □

The same construction as in the previous proof allows us to reduce the minimality problem to the hyper-minimality problem, so we obtain the following result.

Theorem 4.4 (Hyper-Minimality Problem). *The problem of deciding for a given deterministic finite automaton A, whether A is hyper-minimal, is* NL-*complete.*

Proof. To prove NL-hardness, we reduce the NL-complete minimality problem for DFAs [22] to the hyper-minimality problem. Given some DFA A over an alphabet Σ, we construct a DFA A' with a new input symbol $\# \notin \Sigma$ on which every state goes back to the initial state. It is easy to see that A' is minimal if and only if A is minimal, because the newly introduced $\#$ symbol cannot be used to distinguish any states. Further, since all states in A' are kernel states, we also know by Theorem 3.1 that A' is hyper-minimal if and only if A' is minimal, which in turn holds if and only if the automaton A is minimal.

It remains to prove containment in NL. We do this by describing a nondeterministic logspace algorithm for the complement of the problem, that is, for deciding whether a given DFA A is *not* hyper-minimal. Since NL = coNL [65, 113], the statement follows. Recall that by Theorem 3.1 a DFA A is not hyper-minimal if and only if A is not minimal, or A has a pair of different but almost-equivalent states, such that one of them is a preamble state. Deciding minimality of automaton A can be done in NL [22]. Further, a nondeterministic logspace bounded Turing machine can guess two distinct states p and q in A, and verify that $p \in \text{Pre}(A)$, and that $p \sim q$. Thus, the problem can be solved in NL. □

Now we turn to the E-minimization problem, which in sharp contrast to the minimization and hyper-minimization problems turns out to be NP-complete, even if the error language E is finite. To prove NP-hardness, it is tempting to use the NP-complete *minimum inferred finite state automaton problem*. This problem is defined in the book of Garey and Johnson [33], where a paper of E. Mark Gold [41] is given as reference. The problem reads as follows:

given finite languages S and T over alphabet Σ, and an integer k, decide whether there exists a k-state DFA A that accepts a language $L \subseteq \Sigma^*$, such that $S \subseteq L$ and $T \subseteq \Sigma^* \setminus L$.

Notice that in a meaningful instance of this problem, the sets S and T must be disjoint. Now the NP lower bound of the E-minimization problem seems to emerge from the following reduction: let A_S be the minimal DFA for the language S, and A_E the minimal DFA for language $E = \Sigma^* \setminus (S \cup T)$—both automata can be constructed easily, because S and T are finite sets. Now there exists a k-state DFA A that is E-equivalent to A_S if and only if this DFA satisfies $S \subseteq L(A)$ and $T \subseteq \Sigma^* \setminus L(A)$.

Unfortunately, Gold [41] defines the problem for Mealey machines, instead of deterministic finite automata as studied here. The trouble with this is quite subtle: a Mealey machine is basically a DFA that does not have accepting states, but rather outputs a symbol whenever reading an input. Assume the output alphabet is $\{0, 1\}$, then we can define the recognized language of a Mealey machine as the set of words, for which the machine's last output is 1. Every such Mealey machine can be converted to an ordinary DFA accepting the same language. Because we may need accepting and non-accepting copies of states, the DFA may need up to twice the number of states as the Mealey machine, and this is a problem for the reduction.

To prove our NP-hardness result, we give a reduction from an NP-complete variant of the satisfiability problem instead, namely *monotone* 3-SAT [33], which is defined as follows:

given a Boolean formula φ in conjunctive normal form where each clause has exactly three positive literals or exactly three negative literals, decide whether φ is satisfiable.

Now we are ready to show the following theorem.

Theorem 4.5 (*E-Minimization Problem*). *The problem of deciding for two given deterministic finite automata A and A_E, and an integer n, whether there exists a deterministic finite automaton B with n states, such that $A \sim_E B$, for $E = L(A_E)$, is* NP*-complete. This even holds if the language E is finite.*

Proof. Since $A \sim_E B$ can be verified for DFAs A and B in deterministic polynomial time by Theorem 4.2, the problem description gives rise to the following guess-and-check algorithm on a nondeterministic polynomial time bounded Turing machine: given the DFAs A and A_E, and integer n, the Turing machine nondeterministically guesses an n-state DFA B and writes it on its working tape. Then it checks whether $A \sim_E B$, for $E = L(A_E)$. If this is the case the machine accepts, otherwise it rejects. Hence the problem belongs to NP.

For proving NP-hardness, we give a reduction from monotone 3-SAT. As an instance of that problem consider a Boolean formula $\varphi = c_0 \wedge c_1 \wedge \cdots \wedge c_{k-1}$ in conjunctive normal form over the set of variables $X = \{x_0, x_1, \ldots, x_{n-1}\}$, where each c_i is either a positive clause of the form $c_i = (x_{i_1} \vee x_{i_2} \vee x_{i_3})$ or a negative clause of the form $c_i = (\overline{x}_{i_1} \vee \overline{x}_{i_2} \vee \overline{x}_{i_3})$. We construct a DFA A and a finite language E as follows. Let $A = (Q_A, \Sigma, \delta, q_0, \{f\})$ be the automaton with state set $Q_A = Q \cup P \cup \{r, f, s\}$, where $Q = \{q_0, q_1, \ldots, q_{k-1}\}$, $P = \{p_0, p_1, \ldots, p_{n-1}\}$,

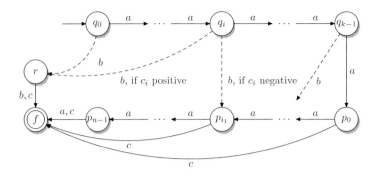

Figure 4.1: The DFA A constructed from the Boolean formula φ with clauses $c_0, c_1, \ldots, c_{k-1}$ and variables $x_0, x_1, \ldots, x_{n-1}$. The transitions on input symbol b from the states $q_0, q_1, \ldots, q_{k-1}$ are only sketched—we have $\delta(q_i, b) = p_{i_1}$ if $c_i = (\overline{x}_{i_1} \vee \overline{x}_{i_2} \vee \overline{x}_{i_3})$, and $\delta(q_i, b) = r$ otherwise. All undefined transitions go to the sink state s, which is not shown.

input alphabet $\Sigma = \{a, b, c\}$, and with the transition function δ which is defined as follows:

$$\delta(q_i, a) = \begin{cases} q_{i+1} & \text{if } 0 \le i \le k-2, \\ p_0 & \text{if } i = k-1, \end{cases} \qquad \delta(p_i, a) = \begin{cases} p_{i+1} & \text{if } 0 \le i \le n-2, \\ f & \text{if } i = n-1, \end{cases}$$

$$\delta(q_i, b) = \begin{cases} r & \text{if } c_i = (x_{i_1} \vee x_{i_2} \vee x_{i_3}), \\ p_{i_1} & \text{if } c_i = (\overline{x}_{i_1} \vee \overline{x}_{i_2} \vee \overline{x}_{i_3}), \end{cases} \qquad \delta(p_i, b) = s, \quad 0 \le i \le n-1,$$

$$\delta(q_i, c) = s, \quad 0 \le i \le k-1, \qquad\qquad \delta(p_i, c) = f, \quad 0 \le i \le n-1,$$

$$\delta(r, a) = s, \ \delta(r, b) = \delta(r, c) = f, \qquad\qquad \delta(f, d) = \delta(s, d) = s, \text{ for } d \in \Sigma.$$

The automaton A is depicted in Figure 4.1. The finite error language is

$$E = \{\, a^i b a^{n-j} \mid 0 \le i \le k-1, \ c_i \text{ contains } x_j \text{ or } \overline{x}_j \,\}$$
$$\cup \{\, a^i b a^j b, a^i b a^j c \mid 0 \le i \le k-1, \ 1 \le j \le n-1 \,\}$$
$$\cup \{\, a^{k+j} b \mid 0 \le j \le n-1 \,\}.$$

A deterministic automaton A_E that accepts the finite language E can easily be constructed. The instance to the E-minimization problem now consists of the DFAs A, and A_E and the integer $k + n + 2$, which is exactly by one smaller than the number of states in A.[1] We show that φ is satisfiable if and only if there exists a DFA B, with $A \sim_E B$, that has at most $k + n + 2$ states, in other words, that has fewer states than A.

---

[1] A tiny example for this construction can be found in Figure 5.5 on page 80, where the automaton A and the error set E are constructed from the Boolean formula $\varphi = x_0$.

First assume that $\alpha\colon X \to \{0,1\}$ is a satisfying assignment for the formula φ. Then there exists a mapping $\chi\colon \{0,1,\ldots,k-1\} \to \{0,1,\ldots,n-1\}$ such that every clause c_i of the formula is satisfied by the variable $x_{\chi(i)}$. We define the DFA $B = (Q \cup P \cup \{f,s\}, \{a,b,c\}, \delta_B, q_0, \{f\})$, where δ_B is defined like δ, except for the b-transitions in states $q_i \in Q$ and $p_i \in P$:

$$\delta_B(q_i, b) = p_{\chi(i)}, \qquad \text{and} \qquad \delta_B(p_i, b) = \begin{cases} f & \text{if } \alpha(x_i) = 1, \\ s & \text{if } \alpha(x_i) = 0. \end{cases}$$

We now prove that B is E-equivalent to A, by showing that all words w from $L(A) \bigtriangleup L(B)$ belong to E. Since, except for input b, the transition functions of A and B are identical, all words $w \in L(A) \bigtriangleup L(B)$ contain at least one b symbol. Further, no word containing more than two b symbols is accepted by A or B, so we only need to consider words that contain exactly one or two b symbols.

First assume $w \in L(A) \setminus L(B)$, and $|w|_b = 2$. Then it must be $w = a^i bb$, where c_i is a positive clause. But then $\delta_B(q_0, a^i bb) = \delta(q_i, bb) = \delta(p_{\chi(i)}, b) = f$, which means $w \in L(B)$—a contradiction to $w \in L(A) \setminus L(B)$.

Now assume $w \in L(A) \setminus L(B)$, and $w = ubv$ for some words $u, v \in \{a,c\}^*$. Since from all states $p_j \in P$ in the DFA A, the input b leads to the sink state, it must be $u = a^i$ with $0 \leq i \leq k-1$, such that b is read from a state $q_i \in Q$. If c_i is a positive clause, then $\delta(q_0, ub) = r$, and then v must be b or c. Then $w \in L(B)$ because $\delta_B(ubv) = \delta_B(q_i, bv) = \delta_B(p_{\chi(i)}, v) = f$, which is clear if $v = c$, and for $v = b$ follows from $\alpha(x_{\chi(i)}) = 1$ since c_i is positive. But $w \in L(B)$ again contradicts our assumption. If otherwise c_i is a negative clause $c_i = (\overline{x}_{i_1} \vee \overline{x}_{i_2} \vee \overline{x}_{i_3})$, then $\delta(q_0, ubv) = \delta(q_i, bv) = \delta(p_{i_1}, v)$, which is accepting if and only if $v = a^{n-i_1}$ or $v = a^j c$ for some $j \in \{1,2,\ldots,n-1-i_1\}$. In both cases we have $ubv \in E$.

Assume $w \in L(B) \setminus L(A)$ and $|w|_b = 2$, then it must be $w = a^i b a^j b$ with $0 \leq i \leq k-1$, and $0 \leq j \leq n-1$. If $j > 0$, then $w \in E$, so let $w = a^i bb \in L(B)$. Then it must be $\delta_B(q_0, a^i bb) = \delta_B(q_i, bb) = \delta_B(q_{\chi(i)}, b) = f$, which means that c_i is a positive clause. But then $\delta(q_0, a^i bb) = \delta(r, b) = f$, so $w \in L(A)$.

Finally assume that $w \in L(B) \setminus L(A)$ and $w = ubv$, for words $u, v \in \{a,c\}^*$. If $\delta_B(q_0, u) = p_i$, for some $i \in \{0,1,\ldots,n-1\}$, then $u = a^{k+i}$ and $v = \lambda$, which means $w \in E$. Thus, let $\delta_B(q_0, u) = q_i$, that is, $u = a^i$ with $0 \leq i \leq k-1$. Then $\delta_B(q_0, a^i bv) = \delta(p_{\chi(i)}, v)$ is accepting if and only if $v = c$, or $v = a^j c$, for some $j \in \{1,2,\ldots,n-1-\chi(i)\}$, or $v = a^{n-\chi(i)}$. The first case $v = c$ leads to a contradiction because $a^i bc \in L(A)$. In the other two cases, it turns out that $w \in E$. Altogether, we obtain $L(A) \bigtriangleup L(B) \subseteq E$.

We have shown that if φ is satisfiable, then there exist a DFA B that is E-equivalent to A, and that has at most $k + n + 2$ states.

For the other direction let $B = (Q_B, \Sigma, \delta_B, q_0^{(B)}, F_B)$ be a DFA with $A \sim_E B$ that has fewer states than A. Since B has to agree with A on acceptance of all words from $\{a,c\}^*$, we can conclude that it has states $Q_B = Q \cup P \cup \{s, f\}$, with $q_0^{(B)} = q_0$, $F_B = \{f\}$, and $\delta_B(t, d) = \delta(t, d)$, for all states $t \in Q_B$ and inputs $d \in \{a, c\}$. Since the words $a^i bc \notin E$, with $0 \leq i \leq k - 1$, are accepted by A, they must also be accepted by B. Thus, for all $i \in \{0, 1, \ldots, k - 1\}$ there is a value $\chi(i) \in \{0, 1, \ldots, n - 1\}$, such that $\delta_B(q_i, b) = p_{\chi(i)}$. Further, all word of the form $a^{k+i} bv \notin E$, with $v \in \{a, b, c\}^+$ and $0 \leq i \leq n - 1$, are not accepted by A, so it must be $\delta_B(p_i, b) \in \{f, s\}$. We define a truth assignment α by $\alpha(x_i) = 1$ if $\delta_B(p_i, b) = f$, and $\alpha(x_i) = 0$ if $\delta_B(p_i, b) = s$. It remains to show that α satisfies φ.

First we show that all positive (negative, respectively) clauses c_i contain the literal $x_{\chi(i)}$ ($\overline{x}_{\chi(i)}$, respectively). Let c_i be a positive clause, and consider the word $a^i ba^{n-\chi(i)}$. Since

$$\delta_B(q_0, a^i ba^{n-\chi(i)}) = \delta_B(q_i, ba^{n-\chi(i)}) = \delta_B(p_{\chi(i)}, a^{n-\chi(i)}) = f,$$

this word is accepted by B, while the word is not accepted by the automaton A, because $\delta(q_0, a^i ba^{n-\chi(i)}) = \delta(q_i, ba^{n-\chi(i)}) = \delta(r, a^{n-\chi(i)}) = s$. Since $A \sim_E B$, it must be $a^i ba^{n-\chi(i)} \in E$, from which follows that c_i contains the literal $x_{\chi(i)}$. Now let $c_i = (\overline{x}_{i_1} \vee \overline{x}_{i_2} \vee \overline{x}_{i_3})$ be negative, and note that $a^i ba^{n-\chi(i)} \in L(B)$ again. If also A accepts $a^i ba^{n-\chi(i)}$, then it must be

$$\delta(q_0, a^i ba^{n-\chi(i)}) = \delta(q_i, ba^{n-\chi(i)}) = \delta(p_{i_1}, a^{n-\chi(i)}) = f,$$

which implies $\chi(i) = i_1$, so the clause c_i contains the literal $\overline{x}_{\chi(i)}$. If A does not accept $a^i ba^{n-\chi(i)}$, then this word must belong to E, which again implies clause c_i contains the literal $\overline{x}_{\chi(i)}$.

Finally we show that for all positive clauses c_i we have $\alpha(x_{\chi(i)}) = 1$ and for all negative clauses c_i we have $\alpha(x_{\chi(i)}) = 0$, which proves the correctness of the reduction. We use words $a^i bb \notin E$, for which A and B must agree on acceptance. If c_i is positive, then $a^i bb$ is accepted by A. Then it must be $\delta(p_{\chi(i)}, b) = f$, which means that $\alpha(x_{\chi(i)}) = 1$. If c_i is negative, then A rejects the word $a^i bb$, so it must be $\delta(p_{\chi(i)}, b) = s$, which means $\alpha(x_{\chi(i)}) = 0$. Thus, every positive (negative) clause c_i is satisfied by the literal $x_{\chi(i)}$ ($\overline{x}_{\chi(i)}$, respectively). So φ is satisfiable if and only if there exists a DFA B, with $A \sim_E B$, that has at most $k + n + 2$ states. $\qquad\square$

Notice that the reduction in the previous proof constructs a DFA A and an error set E such that the automaton A is E-minimal if and only if the given Boolean formula φ is not satisfiable. This immediately gives the following result.

Theorem 4.6 (E-Minimality Problem). *The problem of deciding for given deterministic finite automata A and A_E, whether A is E-minimal, for $E = L(A_E)$, is* coNP*-complete.*

Proof. The coNP-hardness result can be shown with the reduction from the proof of Theorem 4.5. Since the complement of the problem under consideration can be decided by a guess-and-check algorithm in NP—guess an automaton B that is smaller than A and check whether $A \sim_E B$—containment of the E-minimality problem in coNP follows. $\qquad\square$

A summary of the results on minimization and minimality problems for different error profiles can be found in Tables 4.2 and 4.3 at the end of the next subsection. There, also the results for the corresponding problems on NFAs are presented.

4.2.2   Minimal Nondeterministic Finite Automata

Now we turn to nondeterministic automata. We will see that the minimization and minimality problems for the different error profiles are all of same complexity, namely, they are PSPACE-complete. Before we come to these results, we prove a useful result on the descriptional complexity of hyper-minimal NFAs, which is allows us to use the extended fooling set technique of Jean-Camille Birget [8] for almost-equivalent automata.

Lemma 4.7. *Let $L \subseteq \Sigma^*$ be a regular language, and let S be an* extended fooling set *for L, that is, $S = \{ (x_i, y_i) \mid 1 \leq i \leq n \}$, such that for $1 \leq i \leq n$ we have $x_i y_i \in L$, and for $1 \leq i, j \leq n$, with $i \neq j$, we have $x_i y_j \notin L$ or $x_j y_i \notin L$. Further, let $L_0 \subseteq \Sigma^*$ be an infinite language satisfying $vw \in L \Longleftrightarrow w \in L$ for every $v \in L_0$ and $w \in \Sigma^*$. Then every nondeterministic finite automaton A with $L(A) \sim L$ needs at least $|S|$ states.*

Proof. Assume there is an NFA $A = (Q, \Sigma, \delta, q_0, F)$ with $k < n = |S|$ states, satisfying $L(A) \sim L$. Since S is an extended fooling set, we have $L_0 \cdot \{x_i y_i\} \subseteq L$, for all $(x_i, y_i) \in S$. This implies $L_0' \cdot \{x_i y_i\} \subseteq L(A)$ for some infinite subset $L_0' \subseteq L_0$, since otherwise $L \not\sim L(A)$. Then, for all $(x_i, y_i) \in S$, there must be a state p_i in A such that $p_i \in \delta(q_0, vx_i)$ and $\delta(p_i, y_i) \cap F \neq \emptyset$ for infinitely many words $v \in L_0'$. Since $k < n$, there have to be integers i and j, with $1 \leq i, j \leq n$ and $i \neq j$, such that $p_i = p_j$. This means that there are infinitely many words $v, v' \in L_0'$ for which $p_i \in \delta(q_0, vx_i) \cap \delta(q_0, v'x_j)$, and $\delta(p_i, y_i) \cap F \neq \emptyset$ and $\delta(p_i, y_j) \cap F \neq \emptyset$. Because S is an extended fooling set, we have $vx_i y_j \notin L$ or $v'x_j y_i \notin L$, which implies $|L(A) \triangle L| = \infty$—a contradiction to $L(A) \sim L$. Thus, the NFA A needs at least $|S|$ states. $\square$

We are now ready to study the complexity of hyper-minimization problems for nondeterministic finite automata. We first show PSPACE-completeness for the DFA-to-NFA hyper-minimization problem. For proving PSPACE-hardness, we give a reduction from the union universality problem, which asks whether for some given deterministic finite automata $A_1, A_2, \ldots, A_k$ over the common alphabet Σ the equality $\bigcup_{i=1}^{k} L(A_i) = \Sigma^*$ holds—see page 42.

Theorem 4.8. *The problem of deciding for a given deterministic finite automaton A and an integer n, whether there exists a nondeterministic finite automaton B with n states, such that $A \sim B$, is PSPACE-complete.*

Proof. For containment of the problem in PSPACE a straightforward guess-and-check algorithm can be applied—recall that $A \sim B$ can be checked for NFAs in PSPACE by Theorem 4.1.

For proving PSPACE-hardness, we use nearly the same construction as given by Jiang and Ravikumar [67], to reduce the union universality problem for DFAs to the problem under consideration. Let the sequence of DFAs $A_1, A_2, \ldots, A_k$ be an instance of the union universality problem, where for $1 \leq i \leq k$ the automaton $A_i = (Q_i, \Sigma, \delta_i, q_{i,1}, F_i)$ has states $Q_i = \{ q_{i,j} \mid 1 \leq j \leq t_i, \}$, and let $L = \bigcup_{i=1}^{k} L(A_i)$. Further, we may assume that $\{\lambda\} \cup \Sigma \subseteq L$, and that all given DFAs A_i are minimal. We want to construct a DFA A' such that $L = \Sigma^*$ if and only if there is an NFA B with $n = 4 + \sum_{i=1}^{k} t_i$ states, such that $A' \sim B$. Instead of describing the DFA A' directly, we present an equivalent NFA A with $n + 1$ states, from which A' can be constructed in polynomial time. Let $A = (Q, \Sigma', \delta, q_0, F)$ be the NFA with states $Q = \bigcup_{i=1}^{k} Q_i \cup \{q_0, p_1, p_2, p_3, f\}$, final states $F = \bigcup_{i=1}^{k} F_i \cup \{q_0, f\}$, and input alphabet $\Sigma' = \Sigma \cup \{\#, c, d, e\} \cup \Gamma$, where $\Gamma = \{ a_i, b_{i,j} \mid 1 \leq i \leq k, \ 1 \leq j \leq t_i \}$. The transition function δ contains all transitions of the given DFAs, which means $\delta(q_{i,j}, a) = \{\delta_i(q_{i,j}, a)\}$, for all $a \in \Sigma$ and integers i and j, with $1 \leq i \leq k$ and $1 \leq j \leq t_i$. Further, for all $a \in \Sigma$, we define

$$\delta(q_0, a) = \bigcup_{i=1}^{k} \{\delta_i(q_{i,1}, a)\} \cup \{f, p_2\},$$

$$\delta(f, a) = \{f\},$$

and

$$\delta(p_2, a) = \delta(p_3, a) = \{p_3\}.$$

The transitions on symbols from Γ are defined for $1 \leq i \leq k$, and $1 \leq j \leq t_i$ as $\delta(q_0, a_i) = \{q_{i,1}\}$, and $\delta(q_{i,j}, b_{i,j}) = \delta(p_3, b_{i,j}) = \{p_1\}$. The only transitions on symbols c, d, and e are $\delta(q_0, c) = \{p_2\}$, $\delta(p_2, d) = \{p_3\}$, and $\delta(p_3, e) = \{p_1\}$, and the transitions on $\#$ are $\delta(f, \#) = \delta(q_{i,j}, \#) = \{q_0\}$, for $1 \leq i \leq k$ and $1 \leq j \leq t_i$. For all other transitions, with $q \in Q$ and $a \in \Sigma'$, let $\delta(q, a) = \emptyset$. Note that our construction differs from that of Jiang and Ravikumar [67] only in the additional $\#$-transitions. The automaton A is sketched in Figure 4.2 on the next page.

The equivalent DFA A' has about the same structure, and can easily be constructed in polynomial time. Here it is helpful to notice that all states of the power-set automaton, that are reachable from state $\{q_0\}$ by words from $\Sigma^{\geq 2}$, are pairwise equivalent. This is because they have the same outgoing transitions on all symbols from $\Sigma' \setminus \Sigma$ and lead to accepting states on each word from Σ^*.

We now prove that $L = \Sigma^*$ if and only if there is a DFA B with at most n states, such that $A \sim B$. Note that if $L = \Sigma^*$, then we can even construct an *equivalent* NFA B with n states by just deleting state f and all corresponding transitions.

Now assume that $L \neq \Sigma^*$, then there must be some word $w = av \notin L$ for some $a \in \Sigma$ and $v \in \Sigma^+$—recall the assumption $\{\lambda\} \cup \Sigma \subseteq L$. Our goal is to construct an extended fooling set for $L(A)$. For each state $q_{i,j}$, with $1 \leq i \leq k$

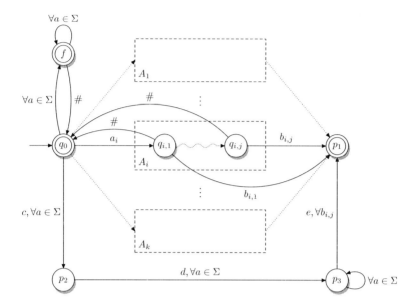

Figure 4.2: The overall structure of the NFA A, which is a slight modification of the NFA constructed by Jiang and Ravikumar [67].

and $1 \leq j \leq t_i$, let $w_{i,j}$ be a word satisfying $\delta_i(q_{i,1}, w_{i,j}) = q_{i,j}$. Using these words, we can construct the following set:

$$S = \{\, (a_i w_{i,j}, b_{i,j}) \mid 1 \leq i \leq k,\ 1 \leq j \leq t_i \,\}$$
$$\cup \{\, (\lambda, cde),\ (c, de),\ (cd, e),\ (cde, \lambda) \,\} \cup \{\, (a, v) \,\}$$

Note that for all $(x, y) \in S$, we have $xy \in L(A)$. Now let (x, y) and (x', y') be two distinct elements from S. In order to show that S is an extended fooling set for $L(A)$ we have to check that $xy' \notin L(A)$ or $x'y \notin L(A)$. We only consider the case where $(x, y) = (a_i w_{i,j}, b_{i,j})$ and $(x', y') = (a, v)$, all other cases can be verified easily. For the sake of contradiction, assume that we have $xy' \in L(A)$, and $x'y \in L(A)$. Then the following holds:

$$xy' \in L(A) \iff a_i w_{i,j} \cdot v \in L(A)$$
$$\iff \delta(q_{i,1}, w_{i,j} v) \cap F \neq \emptyset$$
$$\iff \delta_i(q_{i,j}, v) \in F_i$$

and

$$x'y \in L(A) \iff (a \cdot b_{i,j} \in L(A)$$
$$\iff \delta(q_0, ab_{i,j}) \cap F \neq \emptyset$$
$$\iff \delta_i(q_{i,1}, a) = q_{i,j}.$$

It follows $\delta_i(q_{i,1}, av) \in F_i$—a contradiction to $w = av \notin L = \bigcup_{i=1}^{k} L(A_i)$. Thus, S is an extended fooling set for $L(A)$, and since $\delta(q_0, a\#) = \{q_0\}$, we can use Lemma 4.7 with $L_0 = (a\#)^*$ to see that any NFA B with $A \sim B$ needs at least $|S| = n + 1$ states. Thus, there exists an n-state NFA B, with $A \sim B$, if and only if $\bigcup_{i=1}^{k} L(A) = \Sigma^*$, which concludes the proof. $\qquad\square$

Now the following result on the NFA-to-NFA hyper-minimization problem is immediate.

Corollary 4.9. *The problem of deciding for a given nondeterministic finite automaton A and an integer n, whether there exists a nondeterministic finite automaton B with n states, such that $A \sim B$, is* PSPACE*-complete.*

Proof. Containment in PSPACE can be seen by a guess-and-check algorithm, and PSPACE-hardness follows from Theorem 4.8. $\qquad\square$

It remains to consider the NFA-to-DFA hyper-minimization. As in the case of classical NFA-to-DFA minimization, the integer n defining the size of the target automaton is given in unary notation. Here, PSPACE-hardness will be shown by a reduction from the PSPACE-complete *universality problem* for NFAs, which asks whether $L(A) = \Sigma^*$, for some given NFA A—see page 41.

Theorem 4.10. *The problem of deciding for a given nondeterministic finite automaton A and an integer n, which is given in unary notation, whether there exists a deterministic finite automaton B with n states, such that $A \sim B$, is* PSPACE*-complete.*

Proof. For the upper bound, note that the integer n is given in unary notation. Thus, we have enough space to implemented a guess-and-check algorithm on a nondeterministic polynomial space-bounded Turing machine.

For the lower bound we give a reduction from the universality problem. Let the automaton $C = (Q, \Sigma, \delta_C, q_0, F)$ be an instance of this problem, and assume $q_0 \in F$, that is, $\lambda \in L(C)$—otherwise C is a negative instance of the universality problem, and it can easily be checked whether $q_0 \notin F$. Now choose a new input symbol $\# \notin \Sigma$, let $s, f \notin Q$ be new states, and define the NFA $A = (Q \cup \{s, f\}, \Sigma \cup \{\#\}, \delta, s, F \cup \{f\})$, where the transition function δ is defined as follows, for all states $q \in Q$, and symbols $a \in \Sigma$:

$$\delta(s, a) = \{q_0\}, \qquad \delta(q, a) = \delta_C(q, a), \qquad \delta(f, a) = \{f\},$$
$$\delta(s, \#) = \{f\}, \qquad \delta(q, \#) = \{q_0\}, \qquad \delta(f, \#) = \{f\}.$$

The automaton A is sketched in Figure 4.3 on the following page. Basically, when starting in state q_0 and ignoring the transitions on $\#$, the automaton accepts language $L(C)$. For the integer in the description of the hyper-minimization problem choose $n = 1$. Clearly, this construction can be done by a logarithmic space bounded Turing machine.

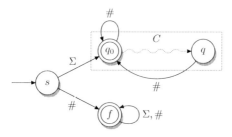

Figure 4.3: The NFA A as instance of the NFA-to-DFA hyper-minimization problem which is constructed from the NFA C.

We will prove the following two statements,[2] from which PSPACE-hardness of the problem under consideration clearly follows.

- If $L(C) \neq \Sigma^*$, then the minimal DFA for the language $L(A)$ is already hyper-minimal and has at least two states.

- If $L(C) = \Sigma^*$, then $L(A) = (\Sigma \cup \{\#\})^+ \sim (\Sigma \cup \{\#\})^*$ and thus, there is a single-state DFA B, with $A \sim B$.

First assume $L(C) \neq \Sigma^*$ and let $A' = (Q', \Sigma \cup \{\#\}, \delta', \{s\}, F')$ be the power-set DFA of A. Note that the only preamble state of A' is the initial state $\{s\}$. We will prove the initial state $\{s\}$ of the DFA A' not to be almost-equivalent to any other state of A'. It then follows that the minimal DFA, that is equivalent to A', is already hyper-minimal. Let $a \in \Sigma$ and $w \in \Sigma^* \setminus L(C)$—recall that $\lambda \in L(C)$. On reading aw from state $\{s\}$, the DFA A' reaches some non-accepting state $P \in Q' \setminus F'$. Further, we have $\delta'(P, \#w) = P$, so the infinite language $aw(\#w)^*$ is not accepted from state $\{s\}$. Since $aw(\#w)^*$ is accepted from state $\{f\}$, these two states cannot be almost-equivalent. Similarly one can see that the infinite language $(\#w)^+$ is accepted from state $\{s\}$, but it is not accepted from any other state $P \in Q'$ with $P \subseteq Q$. Thus, state $\{s\}$ is not almost-equivalent to any other state and so the minimal DFA accepting $L(A)$ is hyper-minimal. Notice that the minimal DFA for $L(A)$ has at least one accepting and one non-accepting state.

Finally, assume $L(C) = \Sigma^*$. Then language accepted by A is $(\Sigma \cup \{\#\})^+$, which is almost-equivalent to the language $(\Sigma \cup \{\#\})^*$, that can be accepted by a single-state DFA. $\qquad\square$

The following theorem summarizes the results on the computational complexity of the DFA-to-NFA, NFA-to-NFA, and NFA-to-DFA hyper-minimization problems.

---

[2] At first glance, these two statements seem a bit over-detailed, but they will be useful later in other proofs, too.

Theorem 4.11 (Hyper-Minimization Problem). *The problem of deciding for a given finite automaton A and an integer n, whether there exists a finite automaton B with n states, such that $A \sim B$, is* PSPACE-*complete, if at least one of the two automata A and B is nondeterministic.*

Proof. The proof is done by Theorems 4.8 and 4.10, and by Corollary 4.9. □

The construction in the proof of Theorem 4.8 can be used to show the following result on the complexity of the hyper-minimality problem.

Theorem 4.12 (Hyper-Minimality Problem). *The problem of deciding for a given nondeterministic finite automaton A, whether A is hyper-minimal, is* PSPACE-*complete.*

Proof. For containment in PSPACE, we apply an easy guess-and-check algorithm for the complement of the problem: given automaton A, a nondeterministic polynomial space-bounded Turing machine guesses an NFA B that is smaller than A and checks whether $A \sim B$. Since NPSPACE = PSPACE [109] and PSPACE = coPSPACE [65, 113], the upper bound follows.

To see that the problem is PSPACE-hard, recall the reduction given in the proof of Theorem 4.8. There an $(n+1)$-state NFA A is constructed from an instance of the union universality problem, such that there exists an n-state NFA B with $A \sim B$ if and only if the instance of the union universality problem is a "yes" instance. In other words, the NFA A is hyper-minimal if and only if the instance of the union universality problem is a "no" instance. □

Now we turn our attention to minimality with respect to E-equivalence, and show that the E-minimization problems for finite automata are PSPACE-complete if nondeterministic automata are involved. Notice that the following result even shows PSPACE-completeness of the DFA-to-DFA E-minimization problem if the error language is specified by a nondeterministic machine.

Theorem 4.13 (E-Minimization Problem). *The problem of deciding for two given finite automata A and A_E, and an integer n,[3] whether there exists a finite automaton B with n states, such that $A \sim_E B$, for $E = L(A_E)$, is* PSPACE-*complete if at least one of the three automata A, B, and A_E is nondeterministic.*

Proof. For containment of the problems in PSPACE we can apply guess-and-check algorithms again—recall that in the case of NFA-to-DFA E-minimization the integer n is given in unary notation.

We now prove PSPACE-hardness. If at least one of the automata A and B in the problem description is nondeterministic, then already the ordinary minimization problems, that is, the NFA-to-NFA, NFA-to-DFA, and DFA-to-NFA minimization problems are PSPACE-hard [67]. Thus, PSPACE-hardness the problems under consideration follows by choosing $E = \emptyset$.

---

[3]Again, in the case where A is an NFA and B is a DFA, the integer n is given in unary notation.

It remains to prove PSPACE-hardness of the DFA-to-DFA E-minimization problem, when E is given by an NFA. Here we can use a reduction from the universality problem for NFAs. Let an NFA C over alphabet Σ be given as an instance of the universality problem. We construct an NFA A_E for the error language $E = L(A_E) = \Sigma \cdot L(C)$ by adding a new initial state to C. Further, let A be the two-state DFA for the language $L(A) = \Sigma^+$, and choose $n = 1$. Clearly this is a logspace computable reduction. Note that $\lambda \notin E$ and $\lambda \notin L(A)$, so the initial state of any DFA that is E-equivalent to A must be non-accepting. For $n = 1$, the only n-state DFA, which could be E-equivalent to A, is the DFA accepting the empty language $\emptyset$, because its single state must not be accepting. Now recall that by Lemma 3.9, we have $A \sim_E B$ if and only if $L(A) \cup E = L(B) \cup E$. Thus, there exists a single-state DFA B with $A \sim_E B$ if and only if $L(A) \cup E = L(B) \cup E = E$. By construction, this equality holds if and only if $E = \Sigma^+$, that is, if and only if $L(C) = \Sigma^*$. This concludes our proof. □

The following result shows that deciding E-minimality is PSPACE-complete if NFAs are involved. Note that even deciding, whether some given DFA is a E-minimal, is PSPACE-complete if the error set is specified by an NFA.

Theorem 4.14 (E-Minimality Problem). *The problem of deciding for two given finite automata A and A_E, whether A is E-minimal, for $E = L(A_E)$, is PSPACE-complete if at least one of the two automata A and A_E is nondeterministic.*

Proof. The PSPACE upper bound follows by easy guess-and-check algorithms for the complement of the problem. For the case, where the automaton A in the problem description may be nondeterministic, the PSPACE lower bound follows from the PSPACE-completeness of the ordinary minimality problem for NFAs [67], by choosing $E = \emptyset$.

It remains to prove PSPACE-hardness for the case where A is a DFA and A_E is an NFA. Here we can use a similar reduction as in the proof of Theorem 4.13: let C be an NFA over the alphabet Σ and construct the DFA A and NFA A_E as in the PSPACE-hardness proof for the DFA-to-DFA E-minimization problem from Theorem 4.13. Note that A has two states, and there exists an E-equivalent single-state DFA B, for $E = L(A_E)$, if and only if $L(C) = \Sigma^*$. Thus, automaton A is not E-minimal if and only if the NFA C is a "no" instance of the universality problem. □

We summarize the complexity of minimization and minimality problems, for NFAs and DFAs, with respect to different error profiles in Tables 4.2 and 4.3.

4.3 Canonicity Problems

In general a hyper-minimal or E-minimal DFA for a language L can be smaller than the minimal DFA that accepts L. But this is not always the case, since

Minimization Problem				
	DFA-to-...		NFA-to-...	
Equivalence relation	DFA	NFA	DFA	NFA
$\equiv$	NL		PSPACE	
$\sim$				
$\sim_E$, for DFA A_E with $E = L(A_E)$	NP			
$\sim_E$, for NFA A_E with $E = L(A_E)$	PSPACE			

Table 4.2: Results on the computational complexity of minimizing finite automata with respect to different equivalence relations.

Minimality Problem		
Equivalence relation	DFA	NFA
$\equiv$	NL	PSPACE
$\sim$		
$\sim_E$, for DFA A_E with $E = L(A_E)$	coNP	
$\sim_E$, for NFA A_E with $E = L(A_E)$	PSPACE	

Table 4.3: Results on the computational complexity of deciding minimality, hyper-minimality, and E-minimality of a given finite automaton.

there are also languages, where a finite or specified set of errors does not allow for a more succinct representation of the language. This leads to the notion of canonical languages as given by Badr et al. [5]: a language L is *canonical* if the minimal DFA accepting L is also hyper-minimal. When using E-minimality instead of hyper-minimality we speak of an *E-canonical* language. Recently, closure properties of the class of canonical languages, as well as descriptional complexity issues of such languages were studied by Andrzej Szepietowski [114]. In this section we study the complexity of the following decision problems. The *canonicity problem* for finite automata is defined as follows:

> given a finite automaton A, decide whether the language $L(A)$ is canonical.

The *E-canonicity problem* is defined similarly:

> given finite automata A and A_E, decide whether the language $L(A)$ is E-canonical, for $E = L(A_E)$.

As in Section 4.2, we first consider DFA problems. The results for the corresponding NFA problems are presented in Subsection 4.3.2. A summary of complexities of the different canonicity problems can be found in Table 4.4 on page 64.

4.3.1 Deterministic Finite Automata

We start our investigations on the canonicity problem for DFAs. For a *minimal* DFA A the question whether $L(A)$ is *not* canonical boils down to verify that A is *not* hyper-minimal, which is the case if it contains a preamble state that is almost-equivalent to some other state. Since we are not necessarily given a minimal DFA as input to the canonicity problem, we need some extra argumentation.

Lemma 4.15. *Let A be some deterministic finite automaton and A' be its equivalent minimal deterministic finite automaton. There is a preamble state p' in A', that is almost-equivalent to some other state r' in A', if and only if there is a preamble state p in A, that is not equivalent to any kernel state q in A, and there is a state r in A that is almost-equivalent but not equivalent to p.*

Proof. First assume there are such states p and r in A. After minimization of A, there must be states p' and r' in A', such that $p' \equiv p$ and $r' \equiv r$. Since $p \not\equiv r$ and $p \sim r$, the states p' and r' must be distinct and almost-equivalent. If p' is in the kernel of A', then there are infinitely many words leading from the initial state of A' to state p'. Then in A, all these words must lead to states that are equivalent to p', so at least one of the states, that are equivalent to p', must be in the kernel of A. But then p is equivalent to this kernel state, which contradicts our assumption. So p' must be a preamble state.

For the other direction, assume that in the minimal DFA A' there is a preamble state p', that is almost-equivalent to some other state r' of A', and let A be some DFA that is equivalent to A'. Then there must be states p and r in A, with $p \equiv p'$, $r \equiv r'$, and of course $p \sim r$ and $p \not\equiv r$. Let $\tilde{p}$ be a state in A that is equivalent to p'. If $\tilde{p}$ is in the kernel of A, then there is an infinite number of words leading from the initial state of A to state $\tilde{p}$. Since A and A' are equivalent, all these words must lead from the initial state of A' to state p'. This is a contradiction to p' being a preamble state, so all states $\tilde{p}$ in A, with $\tilde{p} \equiv p'$, are preamble states. Thus, in A there are almost-equivalent states p and r, where p is a preamble state that is not equivalent to any kernel state. $\qquad \square$

For the proof of our next theorem, we use a reduction from the *reachability problem* for directed graphs of out-degree two:

> given a directed graph $G = (V, E)$, where each vertex has at most two successors, and two vertices $s, t \in V$, decide whether there is a path from s to t in G.

This problem is known to be NL-complete [70], even for acyclic graphs.

Theorem 4.16 (Canonicity Problem). *The problem of deciding for a given deterministic finite automaton A, whether language $L(A)$ is canonical, is NL-complete.*

Proof. We first prove containment of the complement of our problem in NL. Since NL = coNL [65, 113], the upper bound follows. By definition, a language is *not* canonical if its minimal DFA is *not* hyper-minimal, and a minimal DFA is *not* hyper-minimal if and only if it contains a preamble state that is almost-equivalent to some other state—cf. Theorem 3.1. We decide this property on a not necessarily minimal DFA A with the help of Lemma 4.15. To this end a nondeterministic Turing machine guesses states p and r of A, and verifies the following four properties in sequence: (i) p is a preamble state, (ii) $p \not\equiv r$, (iii) $p \sim r$, and (iv) $p \not\equiv q$, for all kernel states q of A. If all answers are positive, the machine accepts, otherwise it rejects. Since all involved properties (and their complements) can be checked in NL, the algorithm can be implemented on a logarithmic space-bounded Turing machine.

It remains to prove NL-hardness, which will be done by a reduction from the above mentioned reachability problem for acyclic graphs. As an instance of that problem, let $G = (V, E)$ be an acyclic graph and $s, t \in V$. We construct a DFA A as follows. The set of states is $V \cup \{f\}$, the initial state is s, the only final state is t, and the transition function δ is defined as follows: whenever a vertex $v \in V \setminus \{t\}$ has two successors v_1 and v_2, then $\delta(v, a) = v_1$ and $\delta(v, b) = v_2$. If there is only one successor, then $\delta(v, b) = f$, and if v has no successor, then $\delta(v, a) = \delta(v, b) = f$. The transitions from states t and f are defined by $\delta(t, a) = \delta(t, b) = \delta(f, a) = \delta(f, b) = f$. If in G there is no path from s to t, then A accepts the empty language, which is a canonical language. If otherwise there is a path from s to t, then the accepted language is not empty but finite. In this case A, and also the equivalent minimal DFA for $L(A)$ have at least two states. But since the language $L(A)$ is finite, it is almost-equivalent to the empty language, which is accepted by a single-state DFA, so $L(A)$ is not canonical. Thus, there is a path from s to t in G if and only if the language $L(A)$ is not canonical. This proves NL-hardness because NL = coNL [65, 113]. $\square$

Similar to the situation of minimization problems for DFAs, where E-equivalence induces a higher complexity than almost-equivalence and classical equivalence, also the E-canonicity problem for DFAs is harder than the canonicity problem, namely coNP-complete.

Theorem 4.17 (*E*-Canonicity Problem). *The problem of deciding for two given deterministic finite automata A and A_E, whether language $L(A)$ is E-canonical, for $E = L(A_E)$, is* coNP*-complete, even if E is a finite language.*

Proof. For the upper bound we show NP-completeness for the complement of the problem under consideration, which is the following: given two DFAs A and A_E, is there a DFA B with fewer states than the minimal DFA for $L(A)$, satisfying $A \sim_E B$, for $E = L(A_E)$? Given the automaton A, we can construct a representation of the minimal DFA A' for $L(A)$ in deterministic polynomial time. Since also $A \sim_E B$ can be verified in deterministic polynomial time by Theorem 4.2, the problem description gives rise to a straightforward guess-and-check algorithm on a nondeterministic polynomial time bounded Turing machine. Hence, deciding whether $L(A)$ is *not* E-canonical can be done in NP.

The lower bound follows from the reduction of the satisfiability problem for the NP-completeness proof of Theorem 4.5. From a given Boolean formula φ, two DFAs A and A_E are constructed such that A is E-minimal, for $E - L(A_E)$, if and only if the formula φ is satisfiable. Further, one can check that the automaton A is a minimal DFA if the given Boolean formula contains at least one positive clause, and this can be assumed because otherwise the formula φ is trivially satisfiable. Thus, the language $L(A)$ is *not* E-canonical if and only if φ is satisfiable. This shows that the E-canonicity problem for DFAs is coNP-hard. □

4.3.2   Nondeterministic Finite Automata

Now we consider the problems of deciding, whether the language accepted by a given nondeterministic finite automaton is canonical, or E-canonical. We start with the former problem.

Theorem 4.18 (Canonicity Problem). *The problem of deciding for a given nondeterministic finite automaton A, whether $L(A)$ is canonical, is* PSPACE-*complete.*

Proof. For the upper bound we combine an on-the-fly power-set construction, similar as in the proof of Theorem 4.1, on the given input NFA with the NL algorithm for the corresponding DFA problem, which results in a PSPACE algorithm.

For the lower bound recall the proof of Theorem 4.10, where a reduction from the PSPACE-complete universality problem is presented. Given an NFA C as an instance of this problem, an NFA A with the following two properties is constructed.

- If $L(C) \neq \Sigma^*$, then the minimal DFA for the language $L(A)$ is already hyper-minimal and has at least two states.

- If $L(C) = \Sigma^*$, then $L(A) = (\Sigma \cup \{\#\})^+ \sim (\Sigma \cup \{\#\})^*$ and thus, there is a single-state DFA B, with $A \sim B$.

The first statement implies that the language $L(A)$ is canonical if $L(C) \neq \Sigma^*$. Further, if $L(C) = \Sigma^*$, then $L(A)$ is *not* canonical, because the minimal DFA for $L(A)$ needs two states, but there is an almost-equivalent single-state DFA B. In conclusion, the language $L(A)$ is *not* canonical if and only if $L(C) = \Sigma^*$. Since PSPACE is closed under complementation, the result follows. □

It remains to consider the E-canonicity problem for nondeterministic machines. In the problem definition of E-canonicity we have two finite automata: the automaton A, the language of which should be checked for E-equivalence, and the automaton A_E specifying the error language E. The case where both given automata are deterministic was handled in Theorem 4.17. The following result considers the remaining cases. The complexity class coNEXP used in the theorem consists of the complements of problems in NEXP, which is the class of problems that are solvable by nondeterministic Turing machines within exponential time. One can see that the presented complexity bounds are not tight, so the exact complexity of the E-canonicity problems is still an open problem.

Theorem 4.19 (E-Canonicity Problem). *The problem of deciding for two given finite automata A and A_E, with $E = L(A_E)$, whether $L(A)$ is an E-canonical language, is*

- PSPACE-*hard and contained in* coNEXP *if A is nondeterministic, regardless whether A_E is deterministic or nondeterministic,*

- coNP-*hard and contained in* PSPACE *if A is deterministic and A_E is nondeterministic.*

Proof. The coNP-hardness of the problem where A is a DFA and A_E is an NFA follows from Theorem 4.17. Containment of this problem in PSPACE can be seen by the following guess-and-check algorithm for the complement of the problem. Given a DFA A and an NFA A_E, we first construct the minimal DFA A' that is equivalent to A. Then we guess some DFA B with less states than A', and check, whether $A' \sim_E B$, for $E = L(A_E)$—this is true if and only if $L(A)$ is *not* E-canonical. In this case we accept, otherwise we reject.

It remains to consider the E-canonicity problem, where the automaton A (and maybe also A_E) is an NFA. For the upper bound we argue as follows. Given NFAs A and A_E, we construct the equivalent power-set DFAs A' and A'_E, and run the coNP algorithm—see Theorem 4.17—for deciding E-canonicity for DFAs on A' and A'_E. This gives a coNEXP algorithm.

For the PSPACE lower bound we can use the same technique as in the proofs of Theorems 4.10 and 4.18. Let the NFA C be an instance of the universality problem over alphabet Σ, and A be the NFA constructed from C as in the mentioned proofs. Further, let A_E be the minimal DFA for the error language $E = \{\lambda\}$. Recall that the following statements hold.

- If $L(C) \neq \Sigma^*$, then the minimal DFA for the language $L(A)$ is already hyper-minimal and has at least two states.

- If $L(C) = \Sigma^*$, then $L(A) = (\Sigma \cup \{\#\})^+ \sim (\Sigma \cup \{\#\})^*$ and thus, there is a single-state DFA B, with $A \sim B$.

Now we argue in the following way: if $L(C) = \Sigma^*$, then $L(A) = (\Sigma \cup \{\#\})^+$ and any NFA needs at least two states to accept $L(A)$. But the language $(\Sigma \cup \{\#\})^*$, which is E-equivalent to $L(A)$, can be accepted by a DFA with a single state. In this case $L(A)$ is not E-canonical. On the other hand, if $L \neq \Sigma^*$, then the minimal DFA for $L(A)$ is already hyper-minimal. In fact, we know that for $E = \{\lambda\}$, the minimal DFA for $L(A)$ is also E-minimal, because any automaton that is E-equivalent to A is also almost-equivalent to A, since E is a finite set. Thus, $L(A)$ is E-canonical in this case. This concludes our proof. □

The complexities of canonicity and E-canonicity problems for finite automata are summarized in Table 4.4. Note that the canonicity problem for the error profile of classical equivalence—the question, whether the minimal DFA is minimal—is trivial.

Canonicity Problem		
Equivalence relation	DFA	NFA
$\equiv$	trivial	
$\sim$	NL	PSPACE
$\sim_E$, for DFA A_E with $E = L(A_E)$	coNP	PSPACE $\leq \cdot$,
$\sim_E$, for NFA A_E with $E = L(A_E)$	coNP $\leq \cdot$, $\cdot \in$ PSPACE	$\cdot \in$ coNEXP

Table 4.4: Results on the computational complexity of deciding canonicity and E-canonicity of languages accepted by given finite automata. For a complexity class C, the notation $\mathsf{C} \leq \cdot$ means that the problem under consideration is hard for C, and $\cdot \in \mathsf{C}$ means that the problem is contained in C.

4.4 Counting Problems

An interesting characteristic of regular languages is that for each regular language there is a unique minimal DFA accepting this language. In other words, given a DFA A, the number of minimal DFAs B, with $A \equiv B$, is equal to 1. Since hyper-minimal and E-minimal DFAs are not necessarily unique anymore, we are led with the following counting problem:

> given a deterministic finite automaton A, compute the number of hyper-minimal DFAs B, with $A \sim B$.

Naturally, this generalizes to determine the number of E-minimal DFAs, and further to NFAs as input. Counting problems for finite automata were previously investigated, for example by Carme Àlvarez and Birgit Jenner [3], Markus Holzer [48], and Richard E. Ladner [80].

Let us shortly describe some important classes of counting problems. Let FP be the class of functions computable by polynomial time bounded Turing machines. Higher counting complexity classes are introduced via a predicate based approach—see the paper of Lane A. Hemaspaandra and Heribert Vollmer [47]. If C is a complexity class of decision problems, let $\# \cdot \mathsf{C}$ be the class of all functions f such that $f(x) = |\{\, y \mid R(x,y) \text{ and } |y| = p(|x|) \,\}|$, for some C-computable two-argument predicate R and some polynomial p. The class $\# \cdot \mathsf{P}$ coincides with the counting class #P, introduced by Leslie G. Valiant [122]. Moreover, we have the inclusion chain

$$\#P = \# \cdot P \subseteq \# \cdot NP \subseteq \# \cdot P^{NP} = \# \cdot coNP,$$

where the last equality is shown by Seinosuke Toda [117]. Here P^{NP} is defined via deterministic polynomial time bounded Turing machines with an NP *oracle*— such a machine is allowed to ask its oracle questions of the form "$w \in L$?", where L is some NP problem, by writing the query w on a special query-tape. The machine then receives the answer in one computation step.

We show that there is again a significant difference between the computational complexities of questions concerning almost- and E-equivalence. Our first goal is to prove that the counting problem for hyper-minimal DFAs lies in FP. For this we first derive the following lemma, which describes the exact number of hyper-minimal DFAs that are almost-equivalent to a given hyper-minimal DFA.

Lemma 4.20. *Let A be a hyper-minimal deterministic finite automaton that has $p = |\mathrm{Pre}(A)|$ preamble states. Further, let $K_1, K_2, \ldots, K_m$ be the almost-equivalence classes in the kernel of A, and for $1 \leq i \leq m$ let p_i be the number of transitions that lead from preamble states to some state in K_i. Then the number of hyper-minimal deterministic finite automata that are almost-equivalent to A is $2^p \cdot \prod_{i=1}^{m} |K_i|^{p_i}$, if $p > 0$, and it is $|K_s|$, if $p = 0$ and the initial state lies in K_s.*

Proof. First assume $p = 0$ and let K_s be the almost-equivalence class containing the initial state. Since two DFAs are almost-equivalent if and only if their initial states are almost-equivalent, we get $|K_s|$ different hyper-minimal almost-equivalent DFAs, by choosing any of the states in K_s as a new initial state. Now let $p > 0$ and let q be some preamble state of A that goes to a kernel state q' on some input a. Consider the DFA B that is derived from A by changing the target state for this a-transition from q' to q'', for some state q'' of A. Then we have $A \sim B$ if and only if $q' \sim q''$, which is seen as follows. The almost-equivalence $q' \sim q''$ means that the symmetric difference $D_{q'q''} = L(q') \bigtriangleup L(q'')$ of the right languages of q' and q'' is finite. Since q is a preamble state, the language L_q, that leads to state q from the initial state of A (and also of B), is also finite. This implies that $L(A) \bigtriangleup L(B) = L_q \cdot a \cdot D_{q'q''}$ is finite, so $A \sim B$. If on the other hand $q' \not\sim q''$, then the difference $D_{q'q''}$ is infinite, which results in $L(A) \bigtriangleup L(B)$ being infinite, so $A \not\sim B$. This means that for each transition from a preamble state q to some kernel state q' we can choose any state q'', with $q' \sim q''$, as the new target state for this transition, and obtain an almost-equivalent DFA. By iterating this process for each transition from a preamble to a kernel state, we get $\prod_{i=1}^{m} |K_i|^{p_i}$ different almost-equivalent DFAs.

Further, we can obtain 2^p copies of any of these DFAs by freely choosing, which of the p preamble states should be accepting or not, which gives $2^p \cdot \prod_{i=1}^{m} |K_i|^{p_i}$ different hyper-minimal DFAs, that are almost-equivalent to A.

Since we know by Theorem 3.3 that two hyper-minimal almost-equivalent DFAs with non-empty preambles can only differ in the acceptance values of preamble states, and in transitions leading from the preamble into the kernel, there is no other almost-equivalent hyper-minimal DFA that cannot be obtained in this fashion. □

Example 4.21. Consider again the two hyper-minimal DFAs A' and A'' from the examples in Section 3.1. For convenience, the automata are depicted in Figure 4.4. Let us apply Lemma 4.20 to DFA A'. The automaton has a single preamble state q_0, so we have $p = 1$. The kernel consists of the $m = 3$ almost-equivalence classes $K_1 = \{q_1'\}$, $K_2 = \{q_4'\}$, and $K_3 = \{q_2', q_5'\}$. For each of these classes there is exactly one transition leading from the preamble to the respective almost-equivalence class, hence $p_1 = p_2 = p_3 = 1$. It follows that the number of hyper-minimal DFAs that are almost-equivalent to A' is

$$2^p \cdot \prod_{i=1}^{m} |K_i|^{p_i} = 2^1 \cdot (1^1 \cdot 1^1 \cdot 2^1) = 4.$$

Besides the two DFAs A' and A'' from Figure 4.4, there are exactly two other hyper-minimal DFAs that are almost-equivalent to these automata. They can be obtained by making the initial state of A' accepting, and by making the initial state of A'' non-accepting. ◇

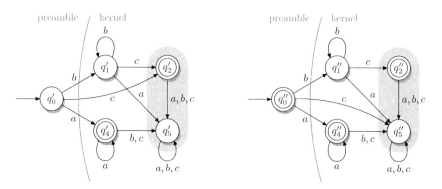

Figure 4.4: The two almost-equivalent hyper-minimal DFAs A' (left) and A'' (right). The gray shading denotes almost-equivalence of states.

Since the values p, $|K_i|$, and p_i, for $1 \le i \le m$, from Lemma 4.20 can be derived from a given DFA in polynomial time, we obtain the following result.

Theorem 4.22 (Counting Hyper-Minimal DFAs). *Given a deterministic finite automaton A, the number of hyper-minimal deterministic finite automata B satisfying $A \sim B$ can be computed in polynomial time, which means that this counting problem belongs to* FP.

Proof. Since hyper-minimization can be done in polynomial time, we can assume that the input DFA is already hyper-minimal. Given a hyper-minimal DFA A with transition function δ, the preamble states can easily be identified by an NL algorithm. In particular p can be computed in polynomial time. Further, the values $|K_i|$ and p_i can be obtained by the following NL algorithm, iterated for each unmarked kernel state q. For all kernel states q', whenever $q \sim q'$, mark q' and increase the counter for $|K_i|$. For all preamble states q'', and all input symbols a, whenever $q \sim \delta(q'', a)$, increase the counter for p_i. Thus, the values $|K_1|, |K_2|, \ldots, |K_m|$ and $p_1, p_2, \ldots, p_m$ can be computed in polynomial time. If $p = 0$, we simply output $|K_s|$, where K_s is the number of states that are almost-equivalent to the initial state. If $p > 0$, we output $2^p \cdot \prod_{i=1}^{m} |K_i|^{p_i}$, which can be done since multiplication and exponentiation of binary numbers can be carried out in polynomial time. □

Our next goal is to show that the counting problem for E-minimal DFAs is at least #P-hard, where we use the following result of Valiant [121], which allows us to compute the coefficients σ_i of a formal power series $f(x) = \sum \sigma_i x^i$ under certain conditions.

Lemma 4.23. *Let n, M, and $\sigma_1, \sigma_2, \ldots, \sigma_n$ be positive integers, with $M > 2$ and $\sigma_i < M$, for $0 \leq i \leq n$, and consider the function $f(x) = \sum_{i=0}^{n} \sigma_i x^i$. If the value $f(x_0)$ is known at some point x_0 with $x_0 \geq M^2$, then the values σ_i, for $0 \leq i \leq n$, can be computed in time polynomial in n.*

In the proof of the upcoming theorem we give a reduction from the *monotone 2-SAT problem*, which is shown to be #P-complete by Valiant [121], and which is defined as follows:

given a Boolean formula $\varphi = c_0 \wedge c_1 \wedge \cdots \wedge c_{k-1}$ in conjunctive normal form over a set of variables $X = \{x_0, x_1, \ldots, x_{n-1}\}$, where each clause c_i contains exactly two positive literals, that is, where $c_i = (x_{i_1} \vee x_{i_2})$, with $x_{i_1}, x_{i_2} \in X$ for $1 \leq i \leq k - 1$, compute the number of satisfying truth assignments for φ.

Note that φ is always satisfiable in this case.

Theorem 4.24 (Counting E-Minimal DFAs). *Given two deterministic finite automata A and A_E, the problem of computing the number of E-minimal deterministic finite automata B satisfying $A \sim_E B$, for $E = L(A_E)$, is #P-hard and can be solved in $\# \cdot \mathsf{coNP}$.*

Proof. To prove #P-hardness, we give a reduction from the monotone 2-SAT problem. Let φ be as defined in the problem description above. To this Boolean formula we apply the construction as described in the proof of Theorem 4.5—see page 48—and obtain a DFA $A = (Q \cup P \cup \{r, f, s\}, \{a, b, c\}, \delta, q_0, \{f\})$ with $Q = \{q_0, q_1, \ldots, q_{k-1}\}$, $P = \{p_0, p_1, \ldots, p_{n-1}\}$, and appropriate transition function δ, as well as a DFA A_E accepting the finite error set E.

Let $\mathcal{A}$ be the set of E-minimal DFAs that are E-equivalent to A. We now discuss the connection of the number of satisfying assignments for φ to the number of automata in $\mathcal{A}$. From the NP-completeness proof for Theorem 4.5 we already know that all DFAs $B \in \mathcal{A}$ have the states $P \cup Q \cup \{f, s\}$, and their transitions on inputs a and c are the same as in A. Further, it must be $\delta_B(q_i, b) \in \{p_j \mid c_i \text{ contains } x_j\}$ for $0 \leq i \leq k-1$, and $\delta_B(p_j, b) \subset \{f, s\}$ for $0 \leq j \leq n-1$, where δ_B is the transition function of B.

We now associate to each satisfying truth assignment α the subset $\mathcal{A}_\alpha$ of all DFAs $B \in \mathcal{A}$, whose transition function δ_B satisfies

$$\delta_B(p_j, a) = \begin{cases} f & \text{if } \alpha(x_j) = 1, \\ s & \text{if } \alpha(x_j) = 0. \end{cases}$$

Since $a^i bb \in L(A) \setminus E$, it must be $\delta_B(q_i, bb) = f$, for $0 \leq i \leq k-1$. This means that $\delta_B(q_i, b) \in \{p_j \mid c_i \text{ contains } x_j, \text{ and } \alpha(x_j) = 1\}$. If for some clause $c_i = (x_{i_1} \vee x_{i_2})$ we have $\alpha(x_{i_1}) \neq \alpha(x_{i_2})$, then there is only one possibility for the transition $\delta_B(q_i, b)$, whereas for the case $\alpha(x_{i_1}) = \alpha(x_{i_2}) = 1$, we may choose between both $\delta_B(q_i, b) = p_{i_1}$, and $\delta_B(q_i, b) = p_{i_2}$. If $t(\alpha)$ denotes the number of clauses c_i, where both literals x_{i_1} and x_{i_2} are satisfied by α, then $|\mathcal{A}_\alpha| = 2^{t(\alpha)}$.

Note that for each two distinct satisfying truth assignments α and α', the sets $\mathcal{A}_\alpha$ and $\mathcal{A}_{\alpha'}$ are disjoint. Further, for each DFA $B \in \mathcal{A}$, there is a satisfying truth assignment α, such that $B \in \mathcal{A}$, which was shown in the NP-completeness proof for Theorem 4.5. Thus, the sets $\mathcal{A}_\alpha$ form a partition of $\mathcal{A}$ and we have

$$|\mathcal{A}| = \sum_{\alpha \models \varphi} |\mathcal{A}_\alpha| = \sum_{\alpha \models \varphi} 2^{t(\alpha)} = \sum_{t=0}^{k} \sigma_t 2^t,$$

where σ_t is the number of satisfying truth assignments, for which in exactly t clauses both literals are satisfied. Note that $\sum_{t=0}^{k} \sigma_t$ is the number of satisfying truth assignments for φ. So the challenge is to extract the values σ_t, which is done with the following trick: instead of constructing A from φ, we construct the DFA from the formula $\varphi^{2n} = \varphi \wedge \varphi \wedge \cdots \wedge \varphi$ ($2n$ times). Then we have

$$|\mathcal{A}| = \sum_{\alpha \models \varphi} 2^{2n \cdot t(\alpha)} = \sum_{t=0}^{k} \sigma_t (2^{2n})^t.$$

Since there are at most $M = 2^n$ truth assignments for the variables of φ, we have $\sigma_t \leq M$, for $0 \leq t \leq k$. Further, for $x_0 = 2^{2n}$, we have $x_0 \geq M^2$, so we can use Lemma 4.23, to compute the values σ_t and also their sum in polynomial time from the value $|\mathcal{A}|$. This shows that the problem of counting the number of E-minimal DFAs that are E-equivalent to a given DFA is #P-hard.

It remains to prove that the problem belongs to $\# \cdot \mathsf{coNP}$. Let A be DFA and $\mathcal{A}$ be the set of E-minimal DFAs that are E-equivalent to A. This set can be characterized as follows: a DFA B belongs to $\mathcal{A}$ if and only if (i) $A \sim_E B$ and (ii) there is *no* DFA with exactly one state less than B that is E-equivalent to B (otherwise B would not be E-minimal). By Theorem 4.2 the former condition can be checked in NL and the latter one can be solved in coNP by Theorem 4.5. Hence, both conditions can be verified by a deterministic polynomial time bounded Turing machine with an NP oracle. Here we use the fact the NP oracle can also be used as a coNP oracle. This results in a P^{NP} Turing machine computing the two-argument predicate R_E that asks whether for two given DFAs A and B the above conditions are satisfied. Therefore, counting the number of DFAs that cause the predicate R_E to be true, for the given first argument A, determines the value $|\mathcal{A}|$. This shows that the problem under consideration belongs to $\# \cdot \mathsf{P}^{\mathsf{NP}}$, which is equal to $\# \cdot \mathsf{coNP}$ [117]. This proves the stated claim. $\qquad\square$

Instead of counting the number of hyper-minimal or E-minimal DFAs, one could also count minimal NFAs for the equivalence relations under consideration, also for classical equivalence. It is easy to see, using a similar strategy as in the proof of Theorem 4.24, that these three NFA counting problems belong to #PSPACE, which is be equal to FPSPACE, the class of functions computable in polynomial space—see the paper of Ladner [80]. What can be said about the lower bound on these NFA counting problems? We have to leave open this question for counting minimal and hyper-minimal NFAs. For counting the number of E-minimal NFAs, we can prove #P-hardness with nearly the same proof as for Theorem 4.24.

Theorem 4.25 (Counting E-Minimal NFAs). *Given a nondeterministic finite automaton A and a deterministic finite automaton A_E, the problem of computing the number of different E-minimal nondeterministic finite automata B, with $A \sim_E B$ and $E = L(A_E)$, is #P-hard and contained in #PSPACE.*

Proof. Counting the number of E-minimal NFAs is easily seen to be solvable in #PSPACE. For the #P-hardness, we can use nearly the same proof as for Theorem 4.24. The structure of the automaton A and the error language E guarantee that any E-minimal NFA B, with $A \sim_E B$ has the *same* transitions on inputs a and c as A has—of course the NFA does not have the non-accepting sink state and corresponding transitions. On reading input b from a state p_i, that corresponds to a clause $c_i = (x_{i_1} \vee x_{i_2})$, the automaton B can only enter some subset of the states $\{q_{i_1}, q_{i_2}\}$—otherwise, some word $a^i b a^{n-j}$ that is *neither* in $L(A)$ *nor* in E would be accepted. In detail, this set $\delta_B(p_i, b)$ may not be empty, since otherwise the word $a^i b c \in L(A)$, which does not belong to E, would not be accepted. Further, if some satisfying truth assignment α satisfies exactly one literal x_{i_ℓ} in c_i, then it must be $\delta_B(p_i, b) = \{q_{i_\ell}\}$, and if both literals x_{i_1} and x_{i_2} are satisfied, then $\delta_B(p_i, b)$ may be any one of the three non-empty subsets of $\{x_{i_1}, x_{i_2}\}$. This means that if a truth assignment α satisfies exactly $t(\alpha)$ clauses twice, and all other clauses once, then there are $3^{t(\alpha)}$ different hyper-minimal NFAs associated to this truth assignment. By constructing the automaton from φ^{2n}, we can again utilize Lemma 4.23 to calculate the number $\sum_{t=0}^{k} \sigma_t$ of satisfying truth assignments for φ from the number $\sum_{t=0}^{k} \sigma_t (3^{2n})^t$ of satisfying truth assignments for φ^{2n}—here σ_t is again the number of satisfying truth assignments, that satisfy exactly t clauses twice. This proves the stated result. $\square$

Chapter 5

Brzozowski-Like Minimization Algorithms

Tтнis chapter is devoted to actual algorithms for lossy compression of deterministic finite automata, which are based on Brzozowski's algorithm for minimizing finite automata [12]. The study of the classical task of DFA minimization was started in the 1950's by Huffman [63, 64] and Moore [96], and many algorithms for solving that problem have been developed since then. Until today the best known worst case running time for minimizing arbitrary n-state DFAs is $O(n \log n)$, which is achieved by Hopcroft's algorithm [61]. Another very prominent algorithm is Brzozowski's minimization algorithm for finite automata [12], which constructs a minimal DFA from a given finite automaton, regardless whether the input automaton is deterministic or nondeterministic. This algorithm first builds the reversal of the given automaton, which in general results in a nondeterministic machine, and then determinizes this automaton. The resulting DFA, which accepts the reversal of the language accepted by the input automaton, then undergoes the same procedure (reversal and power-set construction) again, which results in the desired minimal DFA. Due to the power-set construction, the worst case running time of Brzozowski's algorithm clearly is exponential. Despite its terrible worst case running time, it is reported by Watson [123] that Brzozowski's minimization algorithm often outperforms Hopcroft's algorithm. This statement was in revised by Marco Almeida, Nelma Moreira, and Rogério Reis [2]: only when starting from a nondeterministic finite automaton, Brzozowski's algorithm seems superior to other algorithms for constructing minimal DFAs.

Besides the (non-)efficiency of the algorithm in practice, Brzozowski's algorithm is interesting from a theoretical point of view. Watson's taxonomy of DFA minimization algorithms [123] classifies algorithms according to different minimization methods, like for example, algorithms that compute equivalence, or inequivalence of states, or use a state merging approach. In this classification, Brzozowski's algorithm is presented isolated from all the other minimization algorithms, because it does not use these specific minimization methods—at least not in an obvious way. But a closer look on the algorithm reveals that it indeed *does* compute state equivalence. A characterization of equivalence of states in a DFA in terms of states of the corresponding reversal DFA is given by Jean-Marc Champarnaud, Ahmed Khorsi, and Thomas Paranthoën [20].

In this chapter we design modifications of Brzozowski's minimization algorithm that can be used to construct deterministic finite automata that are minimal with respect to different error profiles. In particular, we present algorithms for constructing deterministic finite automata that are hyper-minimal (Section 5.3), cover-minimal (Section 5.4), and k-minimal (Section 5.5). The presented algorithms are based on results concerning the more general error profile of E-equivalence, which will be laid out in Sections 5.1 and 5.2. In the latter section we also show that there can be no algorithm that E-minimizes, for a specific error set E, a given finite automaton by merging states. Before we come to this, we fix some notation and shortly recall Brzozowski's minimization algorithm.

Brzozowski's Minimization Algorithm

Brzozowski's algorithm [12] is based on the following two basic constructions for finite automata: the reversal automaton, and the power-set automaton. Let $A = (Q, \Sigma, \delta, I, F)$ be an NNFA. The *reversal automaton*[1] of A is the NNFA A^R that is obtained by reversing all transitions and interchanging initial and final states in A. More formally, define $A^R = (Q, \Sigma, \delta^R, I^R, F^R)$ with the set of initial states $I^R = F$, set of final states $F^R = I$, and the transition function δ^R, which is defined such that

$$p \in \delta^R(q, a) \quad \Longleftrightarrow \quad q \in \delta(p, a),$$

for all states $p, q \in Q$ and symbols $a \in \Sigma$. It is well known that $L(A^R) = L(A)^R$.

The construction of the power-set automaton was already described in Section 2.3. Nevertheless we recall the construction here, with a small modification. For an NNFA $A = (Q, \Sigma, \delta, I, F)$ we define its *power-set automaton* to be the DFA $\mathcal{P}(A) = (Q', \Sigma, \delta', q'_0, F')$, where state set $Q' \subseteq 2^Q$ consists of those subsets of Q that are *reachable* in $\mathcal{P}(A)$ from the initial state $q'_0 = I$, the transition function δ' is defined for all states $P \in Q'$ and symbols $a \in \Sigma$ as

$$\delta'(P, a) = \bigcup_{p \in P} \delta(p, a),$$

and the set of final states is $F' = \{ P \in Q' \mid P \cap F \neq \emptyset \}$. This very definition of $\mathcal{P}(A)$ differs from the one given in Section 2.3 in restricting the state set to only those states, which are reachable from the initial state.

---

[1] In the literature, this automaton is also called the *dual* automaton. Since the language accepted by this automaton is the reversal of the original language, we prefer the name *reversal automaton*. In Section 9.2 of Part III we study a similar construction for biautomata, but there the constructed automaton does not accept the reversal of the original language, which is why we will then use the name dual automaton instead.

Later we also use the following notation. Let $A = (Q, \Sigma, \delta, q_0, F)$ be a deterministic finite automaton and $E \subseteq \Sigma^*$ be an error language. We denote by $Q_A(E)$ the set of states of automaton A that are reachable from q_0 *only* by reading words from the set E. More formally,

$$Q_A(E) = \{ q \in Q \mid L_{A,q} \subseteq E \}.$$

Now Brzozowski's algorithm [12] can simply be described as follows.

Theorem 5.1 (Brzozowski's Minimization Algorithm)*. If A is a deterministic or nondeterministic finite automaton, then $A' = \mathcal{P}([\mathcal{P}(A^R)]^R)$ is the minimal deterministic finite automaton for $L(A)$.*

We illustrate the construction in the following example, which shows that the automaton $\mathcal{P}(A^R)$ plays a central role—not only literally—for the minimality of the automaton $\mathcal{P}([\mathcal{P}(A^R)]^R)$.

Example 5.2. Consider the deterministic finite automaton $A = (Q, \Sigma, \delta, q_0, F)$ with state set $Q = \{q_0, q_1, q_2, q_3, q_4\}$, input alphabet $\Sigma = \{a, b\}$, initial state q_0, final states $F = \{q_1, q_3, q_4\}$, and the transition function δ, which can be read from Figure 5.1.

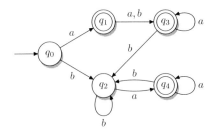

Figure 5.1: The DFA A as input to Brzozowski's minimization algorithm.

In the first step of the algorithm we compute the reversal automaton of A, which result in the NNFA $A^R = (Q, \Sigma, \delta^R, F, \{q_0\})$, with the following transitions:

$$\delta^R(q_0, a) = \emptyset, \qquad \delta^R(q_0, b) = \emptyset,$$
$$\delta^R(q_1, a) = \{q_0\}, \qquad \delta^R(q_1, b) = \emptyset,$$
$$\delta^R(q_2, a) = \emptyset, \qquad \delta^R(q_2, b) = \{q_0, q_2, q_3, q_4\},$$
$$\delta^R(q_3, a) = \{q_1, q_3\}, \qquad \delta^R(q_3, b) = \{q_1\},$$
$$\delta^R(q_4, a) = \{q_2, q_4\}, \qquad \delta^R(q_4, b) = \emptyset.$$

Next, we have to compute the power-set automaton of A^R, which yields the deterministic finite automaton $B = \mathcal{P}(A^R)$. This automaton is depicted in

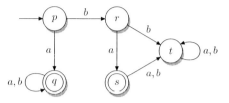

Figure 5.2: The DFA $B = \mathcal{P}(A^R)$ in the intermediate step of Brzozowski's minimization algorithm.

Figure 5.2, where the state labels p, q, r, s, t are abbreviations for the following subsets of Q:

$$p = \{q_1, q_3, q_4\}, \quad q = \{q_0, q_1, q_2, q_3, q_4\} \quad r = \{q_1\}, \quad s = \{q_0\}, \quad t = \emptyset.$$

Reversing the automaton B yields the automaton B^R, and the power-set automaton thereof is the automaton $A' = \mathcal{P}(B^R)$, which is shown in Figure 5.3. Here the state labels stand for the corresponding set, for example the label qs denotes the state $\{q, s\}$ of the automaton A'.

By Theorem 5.1 we know that A' is the minimal DFA for the language $L(A)$. Recall that a DFA is minimal if and only if all its states are reachable, and no two distinct states are equivalent. We have restricted the power-set construction to produce only reachable states, so in order to verify that A' is minimal we need show that all its states are pairwise distinguishable. We do this exemplarily for the two states $\{q, s\}$ and $\{q\}$: the word ab is accepted when starting from state $\{q, s\}$, but it is not accepted when starting from state $\{q\}$. Therefore these two states are not equivalent. What is interesting about the word ab and the sets $\{q, s\}$ and $\{q\}$ is the following. In the automaton B, from which automaton A' is constructed by using reversal and power-set construction, the reversal of the word ab, that is, the word $(ab)^R = ba$ leads from the initial state of B to state s. Further, this state s is an element of the state $\{q, s\}$ of automaton A', but it is not contained in state $\{q\}$. This is why the word ab leads in A' from state $\{q, s\}$ to a final state, but does not lead to a final from state $\{q\}$.

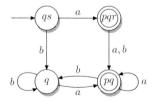

Figure 5.3: The minimal DFA $A' = \mathcal{P}(B^R)$ which is the output of Brzozowski's minimization algorithm.

So the elements, in which two states S and T of $A' = \mathcal{P}(B^R)$ differ, tell us, how to distinguish these two states: simply take the reversal of a word that leads in automaton B to a state which is contained in $S \triangle T$. This observation will be generalized in the following section. $\diamond$

5.1 Structural Characterization of E-Equivalent States

In order to extend Brzozowski's minimization algorithm for other error profiles, we first take a closer look on the power-set automaton $\mathcal{P}(A^R)$ of the reverse of a DFA A. Our goal is to decide for two given states S and T of $\mathcal{P}(A^R)$, and an error language E, whether $S \sim_E T$ holds. The following lemma describes the connection between words accepted from states in $\mathcal{P}(A^R)$, and words leading to states in A. The lemma also holds if the power-set automaton $B = \mathcal{P}(A^R)$ is defined such that $Q_B = 2^Q$, even if not all subsets of Q are reachable in B.

Lemma 5.3. *Let $A = (Q, \Sigma, \delta, q_0, F)$ be a deterministic finite automaton and let $B = (Q_B, \Sigma, \delta_B, F, F_B)$ be the power-set automaton of the reverse of A, that is, $B = \mathcal{P}(A^R)$. Then*

$$L_B(S) = \bigcup_{q \in S} L_{A,q}^R,$$

for all states $S \in Q_B$, and the union is disjoint. Further, for all states S and T of B, we have

$$L(S) \triangle L(T) = \bigcup_{q \in S \triangle T} L_q^R.$$

Proof. For all states $S \in Q_B$ and words $w \in \Sigma^*$, the following holds:

$$
\begin{aligned}
w \in L_B(S) &\Longleftrightarrow \delta_B(S, w) \in F_B \\
&\Longleftrightarrow q_0 \in \delta_B(S, w) \\
&\Longleftrightarrow q_0 \in \delta^R(q, w), \text{ for some } q \in S \\
&\Longleftrightarrow \delta(q_0, w^R) = q, \text{ for some } q \in S \\
&\Longleftrightarrow w^R \in \bigcup_{q \in S} L_{A,q}.
\end{aligned}
$$

This proves the first equation of the statement. Further, since A is a DFA, the left languages $L(A_p)$ and $L(A_q)$ are disjoint, if $p \neq q$, since no word can lead to two distinct states. Now the second equation can be seen as follows:

$$L_B(S) \triangle L_B(T) = \left(\bigcup_{q \in S} L_{A,q}^R \right) \triangle \left(\bigcup_{q \in T} L_{A,q}^R \right) = \bigcup_{q \in S \triangle T} L_{A,q}^R. \qquad \square$$

From this lemma one can easily conclude the following statement, which can also be used to prove Theorem 5.1.

Corollary 5.4. Let $A = (Q, \Sigma, \delta, q_0, F)$ be a deterministic finite automaton where all states are reachable, that is, $L_{A,q} \neq \emptyset$, for every $q \in Q$. Then $\mathcal{P}(A^R)$ is a minimal deterministic finite automaton.

Proof. Let $B = \mathcal{P}(A^R)$. Since we assume all states in B to be reachable because it is constructed by the power-set construction, it remains to show that all states in B are pairwise distinguishable. Therefor let S and T be two distinct states of B. From Lemma 5.3 we get

$$L_B(S) \triangle L_B(T) = \bigcup_{q \in S \triangle T} L_{A,q}^R.$$

Note that there is an element $q \in S \triangle T$, and since all states in A are reachable, the language $L_{A,q}$ is not empty. It follows that also the symmetric difference $L_B(S) \triangle L_B(T)$ is not empty, so S and T are distinguishable. $\square$

Corollary 5.4 implies that $\mathcal{P}([\mathcal{P}(A^R)]^R)$ is a minimal DFA, because the intermediate automaton $\mathcal{P}(A^R)$ has no unreachable states. If we leave out the precondition in Corollary 5.4, that all states in A are reachable, then the resulting DFA is not necessarily minimal, because it may contain distinct states S and T which are not distinguishable. In this case the symmetric difference

$$L_B(S) \triangle L_B(T) = \bigcup_{q \in S \triangle T} L_{A,q}^R$$

of the sets of words accepted from states S and T must be empty, which means that all elements $q \in S \triangle T$ satisfy $L(A_q)^R = \emptyset$. In other words, distinct but equivalent states S and T may only differ in elements q which are non-reachable states of A. In this way, we can *syntactically check*, whether or not two states are equivalent: if they differ in an element that is reachable in A, then they are not equivalent, otherwise they are equivalent. This idea can be adapted also to the general notion of E-equivalence. The next statement is the main lemma of this section.

Lemma 5.5. Let A be a deterministic finite automaton with state set Q and input alphabet Σ. Further, assume $E \subseteq \Sigma^*$, and let S and T be two states of $B = \mathcal{P}(A^R)$. Then $S \sim_E T$ if and only if $S \triangle T \subseteq Q_A(E^R)$, that is, if and only if $L_{A,q} \subseteq E^R$, for all $q \in S \triangle T$.

Proof. By definition of $\sim_E$, and Lemma 5.3, $S \sim_E T$ holds if and only if we have

$$L_B(S) \triangle L_B(T) = \bigcup_{q \in S \triangle T} L_{A,q}^R \subseteq E.$$

So, $S \sim_E T$ if and only if $L_{A,q} \subseteq E^R$, for all $q \in S \triangle T$. $\square$

Since classical equivalence allows no errors, two states S and T are equivalent if they are E-equivalent for $E = \emptyset$, that is, if $S \sim_\emptyset T$. Then Lemma 5.5 states that S and T are equivalent, if and only if $L_{A,q} \subseteq \emptyset^R = \emptyset$, which means that state q is not reachable in A, for all elements $q \in S \triangle T$. Since Brzozowski's minimization algorithm assumes that the power-set construction does not produce unreachable states, it immediately follows that no two different states S and T in the resulting DFA can be equivalent. Lemma 5.5 can also be used to decide equivalence with respect to other error profiles. Consider for example the almost-equivalence relation. Recall that by Theorem 3.12 two states in an n-state DFA are almost-equivalent if and only if they are E-equivalent, for $E = \Sigma^{\leq n-2}$. Then two states S and T of the DFA $B = \mathcal{P}(A^R)$ are almost-equivalent if $S \triangle T$ consists only of states from A, which are only reachable in A by words from $E^R = \Sigma^{\leq n-2}$. In other words, states S and T are almost-equivalent if and only if $S \triangle T \subseteq \mathrm{Pre}(A)$. Table 5.1 summarizes the rules for deciding state equivalence with respect to different error profiles. The relations $\approx_\ell$ and $\sim_k$, as well as the functions level and in-level will be defined in Sections 5.4 and 5.5, respectively.

Relation	Error Language	Condition
$S \sim_E T$	E	$\forall q \in S \triangle T : q \in Q_A(E^R)$
$S \equiv T$	$E = \emptyset$	$\forall q \in S \triangle T : q$ unreachable in A
$S \sim T$	$E = \Sigma^{\leq n-2}$	$\forall q \in S \triangle T : q \in \mathrm{Pre}(A)$
$S \approx_\ell T$	$E = \Sigma^{> \ell - m}$	$\forall q \in S \triangle T : \mathrm{level}_A(q) > \ell - m$
$S \sim_k T$	$E = \Sigma^{< k - m'}$	$\forall q \in S \triangle T : \mathrm{in\text{-}level}_A(q) < k - m'$

Table 5.1: Deciding E-equivalence of states S and T of $\mathcal{P}(A^R)$—cf. Lemma 5.5. In the definition of the error language for the relation $\sim$, the integer n is the number of states of the DFA A. In the definition of the error language for the relation $\approx_\ell$, the number ℓ is the cover length, and $m = \max(\mathrm{level}(S), \mathrm{level}(T))$. Finally, for the relation $\sim_k$, we have $m' = \min(k, \mathrm{in\text{-}level}(S), \mathrm{in\text{-}level}(T))$.

5.2 E-Mergeability of States

In the previous section we found a tool to easily check state equivalence for different error profiles. Since many minimization algorithms rely on state merging, the next step is to investigate, under which conditions two states can be merged. For a DFA $A = (Q, \Sigma, \delta, q_0, F)$, two of its states p and q, and an error language $E \subseteq \Sigma^*$, we say that p is E-mergeable to q, if $A \sim_E A'$, where A' results from A by merging state p to state q—see Section 2.3: we have $A' = (Q \setminus \{p\}, \Sigma, \delta', q_0', F \setminus \{p\})$, with

$$\delta'(r, a) = \begin{cases} q & \text{if } \delta(r, a) = p, \\ \delta(r, a) & \text{otherwise,} \end{cases} \quad \text{and} \quad q_0' = \begin{cases} q & \text{if } q_0 = p, \\ q_0 & \text{otherwise.} \end{cases}$$

The following result characterizes E-mergeability of states in terms of languages related to reachability between these states. Here we use the following notation: for a DFA $A = (Q, \Sigma, \delta, q_0, F)$, a state $r \in Q$, sets of states $S, T \subseteq Q$, and language $L = L(A)$, let $_rL_S^T$ denote the set of words that lead from state r to some state $s \in S$, while only passing through states from T in between. If $S = \{s\}$ is a singleton set, we omit the set braces, and write $_qL_s^T$ instead of $_qL_{\{s\}}^T$. Further, for two languages L and L' let

$$L^{-1}L' = \bigcap_{u \in L} u^{-1}L'.$$

Then the characterization of E-mergeable states reads as follows.

Theorem 5.6. *In a deterministic finite automaton A, with $L(A) = L$, state p is E-mergeable to state q if and only if*

$$[_pL_F^Q] \triangle [(_qL_p^{Q\setminus\{p\}})^* \cdot {}_qL_F^{Q\setminus\{p\}}] \subseteq (_{q_0}L_p^{Q\setminus\{p\}})^{-1}E. \tag{5.1}$$

Proof. Let A' be the DFA that results from merging p to q—the DFAs A and A' are sketched in Figure 5.4. We will show that $A' \sim_E A$ is equivalent to Equation (5.1). First note that

$$L(A') = (L(A) \setminus M) \cup N,$$

where M is the set of words accepted by A, for which the computation in A leads through state p, and N is the set of words that are accepted by A', and that have a prefix that leads to state p in A (and to state q in A'). Formally, we have:

$$M = {}_{q_0}L_p^{Q\setminus\{p\}} \cdot {}_pL_F^Q, \qquad N = {}_{q_0}L_p^{Q\setminus\{p\}} \cdot (_qL_p^{Q\setminus\{p\}})^* \cdot {}_qL_F^{Q\setminus\{p\}}.$$

Now the symmetric difference $L(A) \triangle L(A')$ can be written as follows:

$$\begin{aligned}
&L(A) \triangle L(A') \\
&= [L(A) \setminus L(A')] \cup [L(A') \setminus L(A)] \\
&= [M \setminus N] \cup [N \setminus L(A)] \\
&= [_{q_0}L_p^{Q\setminus\{p\}} \cdot (_pL_F^Q \setminus ((_qL_p^{Q\setminus\{p\}})^* \cdot {}_qL_F^{Q\setminus\{p\}}))] \\
&\quad \cup [_{q_0}L_p^{Q\setminus\{p\}} \cdot ((_qL_p^{Q\setminus\{p\}})^* \cdot {}_qL_F^{Q\setminus\{p\}} \setminus (_{q_0}L_p^{Q\setminus\{p\}})^{-1}L(A))]
\end{aligned} \tag{5.2}$$

In the following we show that the set from Equation (5.2) is contained in the error set E if and only if the inclusion (5.1) holds. First assume

$$[_{q_0}L_p^{Q\setminus\{p\}} \cdot (_pL_F^Q \setminus ((_qL_p^{Q\setminus\{p\}})^* \cdot {}_qL_F^{Q\setminus\{p\}}))] \subseteq E.$$

In other words $[u \cdot (_pL_F^Q \setminus ((_qL_p^{Q\setminus\{p\}})^* \cdot {}_qL_F^{Q\setminus\{p\}}))] \subseteq E$, for every $u \in {}_{q_0}L_p^{Q\setminus\{p\}}$, which in turn implies $[_pL_F^Q \setminus ((_qL_p^{Q\setminus\{p\}})^* \cdot {}_qL_F^{Q\setminus\{p\}})] \subseteq u^{-1}E$ for every word $u \in {}_{q_0}L_p^{Q\setminus\{p\}}$. Therefore we obtain

$$[_pL_F^Q \setminus ((_qL_p^{Q\setminus\{p\}})^* \cdot {}_qL_F^{Q\setminus\{p\}})] \subseteq (_{q_0}L_p^{Q\setminus\{p\}})^{-1}E.$$

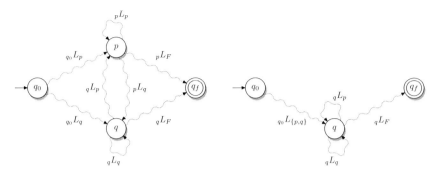

Figure 5.4: Left: the DFA A with states p and q. Right: the DFA A', that results from merging state p to q in DFA A. The depicted languages are to be interpreted such that the corresponding computation paths do not lead through state p or q in between, for example $_qL_p$ means $_qL_p^{Q\setminus\{p,q\}}$—we omit the superscripts for better readability.

By a similar argument starting with the assumption

$$[_{q_0}L_p^{Q\setminus\{p\}} \cdot ((_qL_p^{Q\setminus\{p\}})^* \cdot {}_qL_F^{Q\setminus\{p\}} \setminus (_{q_0}L_p^{Q\setminus\{p\}})^{-1}L(A))] \subseteq E$$

and with the equality $(_{q_0}L_p^{Q\setminus\{p\}})^{-1}L(A) = {}_pL_F^Q$, we find that

$$[((_qL_p^{Q\setminus\{p\}})^* \cdot {}_qL_F^{Q\setminus\{p\}}) \setminus {}_pL_F^Q] \subseteq (_{q_0}L_p^{Q\setminus\{p\}})^{-1}E.$$

We have shown that if the set from Equation (5.2) is contained in E, then

$$[_pL_F^Q \setminus ((_qL_p^{Q\setminus\{p\}})^* \cdot {}_qL_F^{Q\setminus\{p\}})] \cup [((_qL_p^{Q\setminus\{p\}})^* \cdot {}_qL_F^{Q\setminus\{p\}}) \setminus {}_pL_F^Q] \subseteq (_{q_0}L_F^{Q\setminus\{p\}})^{-1}E.$$

Notice that the left-hand side of this equation is the symmetric difference between the languages $_pL_F^Q$ and $(_qL_p^{Q\setminus\{p\}})^* \cdot {}_qL_F^{Q\setminus\{p\}}$. Hence, containment of $L(A) \triangle L(A')$ in E implies Equation (5.1). For the reverse implication, we observe that the language

$$_{q_0}L_p^{Q\setminus\{p\}} \cdot ((_{q_0}L_p^{Q\setminus\{p\}})^{-1}E)$$

is a subset of E. Then Equation (5.1) also implies $L(A) \triangle L(A') \subseteq E$, which gives the stated result. □

The next theorem shows that the simple state merging approach does not work for the problem of E-minimization. Remember that we have already seen in Section 4.2 that the E-minimization problem is NP-complete.

Theorem 5.7. *There is no algorithm that computes, for every two given deterministic finite automata A and A_E, with $E = L(A)$, an E-minimal deterministic finite automaton B for the language $L(A)$ by merging states of A. This even holds for a fixed and finite set E.*

Proof. Consider the DFA A depicted on the left in Figure 5.5, which accepts the language $L(A) = \{aa, ac, bb, bc\}$. If we choose the error language $E = \{ab, ba\}$, then the DFA A', which is depicted on the right in Figure 5.5, is E-minimal and E-equivalent to A.[2]

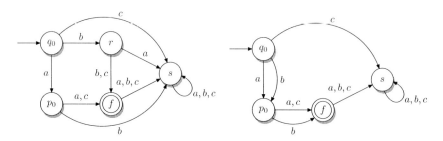

Figure 5.5: Left: the DFA A for the language $L(A) = \{aa, ac, bb, bc\}$. Right: the E-minimal DFA A', for the error set $E = \{ab, ba\}$, accepting the language $L(A') = L(A) \cup E$.

We show that no state in A is E-mergeable to any other state of A. It then follows that A' cannot result from any state merging in A. First consider the initial state q_0: if q_0 is merged to p_0 or to r, then the resulting automaton accepts words of length one, if it is merged to f, then λ is accepted, and if it is merged to s, then no word is accepted at all. Further, if any other state is merged to q_0, then an infinite number of words is accepted, but both $L(A)$ and E are finite, so the resulting automaton cannot be E-equivalent to A. The same argument holds, if state s is merged to some other state. Moreover, we cannot merge state p_0 or state r to s, because then we do not accept the words aa, or bb. State f cannot be merged to any other state, because then no word would be accepted at all, so only states r and p_0 are left to be considered: if r was merged to p_0, then the word bb would not be accepted, and if p_0 would be merged to r, then the word aa would not be accepted. Thus, no state of A can be merged to some other state without introducing errors that do not belong to E. □

Despite this rather negative result, nevertheless for some specific error profiles minimization can be done by state merging algorithms. We will see such algorithms for the error profiles of almost-equivalence, k-equivalence, and cover automata in the upcoming sections.

---

[2]The automaton A and the language E are constructed from the Boolean formula $\varphi = x_0$ as described in the reduction in the proof of Theorem 4.5. Recall that this construction is such that there is a one-to-one correspondence between E-minimal DFAs for A and satisfying truth assignments for φ—see also the proof of Theorem 4.24. Since in our example $\varphi = x_0$ has only one satisfying truth assignment, we know that A' is the unique E-minimal DFA for A.

5.3 Hyper-Minimization

In this section we show how Brzozowski's minimization algorithm can be extended such that it computes from a given finite automaton A an almost-equivalent hyper-minimal DFA A'. Basically we construct the deterministic automaton $\mathcal{P}([\mathcal{P}(A^R)]^R)$ and compute the almost-equivalence classes of that automaton on-the-fly, using Lemma 5.5 from the previous section. Then we merge almost-equivalent states to hyper-minimize the automaton. Recall that by Theorem 3.12 two states of an n-state DFA with input alphabet Σ are almost-equivalent if and only if they are E-equivalent, for $E = \Sigma^{\leq n-2}$. Thus, Lemma 5.5 implies that two states S and T of $\mathcal{P}(B^R)$, for some n-state DFA B with state set Q, are almost-equivalent if and only if

$$S \triangle T \subseteq Q_B(\Sigma^{\leq n-2}) = \mathrm{Pre}(B),$$

which means that the symmetric difference $S \triangle T$ consists only of preamble states of B. So if there is an element $q \in S \triangle T$ that is a kernel state of B, then S and T are not almost-equivalent, and *vice versa*. We summarize this in the following lemma—see also Table 5.1 on page 77.

Lemma 5.8. *Let B be an n-state deterministic finite automaton with input alphabet Σ and state set Q, and let S and T be two states of $\mathcal{P}(B^R)$. Then $S \sim T$ if and only if $S \triangle T \subseteq Q_B(\Sigma^{\leq n-2}) = \mathrm{Pre}(B)$, that is, if and only if all states in $S \triangle T$ are preamble states of B.* $\square$

So if we use Brzozowski's algorithm on some finite automaton A, we can first build the DFA $B = \mathcal{P}(A^R)$, now mark the preamble and kernel states of B, and then continue by constructing the (minimal) DFA $A' = \mathcal{P}(B^R)$. After that, we can use Lemma 5.8 to check which states in A' are almost-equivalent and use state merging to hyper-minimize the automaton. This gives rise to Algorithm 1 for hyper-minimizing finite automata.

Algorithm 1 Brzozowski-like algorithm for hyper-minimizing finite automata.

Require: a DFA or NFA $A = (Q, \Sigma, \delta, q_0, F)$

1: construct $B = \mathcal{P}(A^R)$
2: identify preamble and kernel states of B
3: construct $A' = \mathcal{P}(B^R)$
4: identify preamble and kernel states of A'
5: compute topological order $\prec$ of preamble states of A'
6: **for all** preamble states S of A' **do**
7:    find $T \neq S$ such that $S \triangle T \subseteq \mathrm{Pre}(B)$, and $T \in \mathrm{Ker}(A')$ or $S \prec T$
8:    **if** T exists **then**
9:      merge S to T
10: **return** A'

Theorem 5.9 (Brzozowski-Like Hyper-Minimization). *Given a (deterministic or nondeterministic) finite automaton A, Algorithm 1 computes a hyper-minimal deterministic finite automaton A' that is almost-equivalent to A.*

Proof. We already know by Theorem 5.1 that the DFA $A' = \mathcal{P}(B)$ in line 3 is the minimal DFA for $L(A)$. Further, when the `for` loop in line 6 is finished, there are no preamble states that are almost-equivalent to some other state. So it follows from the characterization of hyper-minimal DFAs in Theorem 3.1 that the resulting DFA is also a hyper-minimal. Since only preamble states S are merged to almost-equivalent states T, where S is not reachable from T, every single merge step preserves almost-equivalence, so A' is almost-equivalent to A. $\qquad\square$

The identification of preamble and kernel states in the automata B and A' can be done on-the-fly during the power-set construction, for example, by using an algorithm of Robert Tarjan [115] to identify strongly connected components. Also the topological ordering of the preamble states of A' can by computed during the second power-set construction. In order to find an appropriate state T in line 7, one could simply cycle through all states of A' and check whether the desired properties are fulfilled. But we do this in a more clever way, by using a hash table H, which gets initialized before the `for all` loop. The entries of H carry states of A' (sets of states of B), and they are indexed by their subset of kernel states of B. Since all states in A' that are equivalent to some state S have the same subset of kernel elements from B, namely $K = S \setminus \mathrm{Pre}(B)$, the entry $H[K] = S$ is a representative of the almost-equivalence class of state S. The initialization of H, which will be described in a few lines, assures that $H[K]$ is always a state to which other almost-equivalent (preamble) states of A' can be merged: preferably $H[K]$ is a kernel state of A', but if there is no kernel state in the almost-equivalence class of $H[K]$, then $H[K]$ is a preamble state that has a maximal index in the topological order $\prec$ among all states that are almost-equivalent to $H[K]$, so that no other almost-equivalent preamble state can be reached from it.

The initialization of H can be done as follows: for all kernel states P of A' we construct the set $K = P \setminus \mathrm{Pre}(B)$, and if $H[K]$ is not yet defined, then we set $H[K] = P$. Then for all preamble states P of A' we construct the set $K = P \setminus \mathrm{Pre}(B)$, and if $H[K]$ is not defined, or if $H[K] = P'$ for some preamble state $P' \prec P$, then we set $H[K] = P$. In this way the entry $H[K]$ is either a kernel state or it is a preamble state that is maximal with respect to the topological order $\prec$. Now the search in line 7 for a state T, to which the preamble state S could be merged, can be done as follows. We compute the set of *critical states* $K = S \setminus \mathrm{Pre}(B)$ and check, whether the entry $H[K]$ in the hash table H is defined. If $H[K]$ is not defined, or if $H[K] = S$, then no appropriate set T is found, otherwise we have $T = H[K]$.

We illustrate the algorithm with the following example.

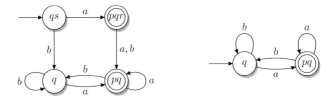

Figure 5.6: Left: the minimal DFA $A' = \mathcal{P}(B^R)$ for the language $L(A)$. Right: the hyper-minimal DFA A_{hypermin} for the language $L(A)$.

Example 5.10. Consider again the DFAs A, $B = \mathcal{P}(A^R)$, and $A' = \mathcal{P}(B^R)$ from Example 5.2. The DFA A' is also depicted on the left-hand side of Figure 5.6. The kernel states in B are q and t, and the preamble states in B are p, r, and s. The kernel states in A' are $\{q\}$ and $\{p, q\}$. If we assume that in the initialization of the hash table H the kernel state $\{q\}$ is processed before the other kernel state $\{p, q\}$, then H has the entry $H[\{q\}] = \{q\}$. Since q is the only element from the two preamble states $\{q, s\}$ and $\{p, q, r\}$ of A', that does not belong to $\mathrm{Pre}(B)$, no further entry is made in H. The topological order of the two preamble states of A' would be $\{q, s\} \prec \{p, q, r\}$.

Assume in the `for all` loop we first process the preamble state $S = \{q, s\}$. We find an appropriate state T by constructing $K = S \setminus \mathrm{Pre}(B) = \{q\}$, and obtain $T = H[K] = \{q\}$. Thus, we merge $\{q, s\}$ to state $\{q\}$—note that this has the effect that the new initial state is now $\{q\}$ instead of $\{q, s\}$. Now we process the preamble state $S' = \{p, q, r\}$: again we obtain $K' = S' \setminus \mathrm{Pre}(B) = \{q\}$, so we also merge state $\{p, q, r\}$ to state $H[K'] = \{q\}$. Since no further preamble states in A' are left, the algorithm terminates, and we obtain the hyper-minimal DFA, which is depicted on the right-hand side of Figure 5.6. $\diamond$

Presumably it is not possible to perform the merge steps on-the-fly while constructing the automaton $A' = \mathcal{P}(B^R)$, because we may only eliminate preamble states, but not kernel states. The problem is that when we discover a state of $\mathcal{P}(B^R)$ for the first time, we may not know, whether it could be a kernel state, unless we already discovered the whole state graph of the automaton $\mathcal{P}(B^R)$.

5.4 Minimal Cover Automata

As a second application of the results from Section 5.1, an algorithm for constructing minimal cover automata is developed, again based on Brzozowski's DFA minimization algorithm [12]. The cover-minimization algorithms given by Câmpeanu et al. [17, 18, 19] and Körner [76] are based on merging similar states. A general study of such similarity relations is given by Jean-Marc Champarnaud, Franck Guingne, and Georges Hansel [21]. Let $A = (Q, \Sigma, \delta, q_0, F)$ be a deterministic finite automaton where all states are reachable. The *level* of a state $q \in Q$, denoted by $\mathrm{level}_A(q)$, is the length of a shortest word leading from q_0 to q in A, that is, $\text{level}_A(q) = \min\{ |w| \mid w \in L_q \}$. Two states p and q are *similar* with respect to cover length $\ell \geq 0$, denoted by $p \approx_\ell q$, if

$$L(p) \cap \Sigma^{\leq \ell - m} = L(q) \cap \Sigma^{\leq \ell - m},$$

where $m = \max\{\text{level}_A(p), \text{level}_A(q)\}$. One readily sees that $p \approx_\ell q$ holds if and only if $p \sim_E q$, for $E = \Sigma^{> \ell - m}$. If we consider, for some DFA B with input alphabet Σ and state set Q, two states S and T of the power-set automaton $\mathcal{P}(B^R)$, Lemma 5.5 implies that S and T are $\Sigma^{>k}$-equivalent if and only if $S \triangle T \subseteq Q_B(\Sigma^{>k})$. This means that $\text{level}_B(q) > k$, for all elements q in the symmetric difference of S and T. Thus, we obtain the following lemma—see Table 5.1 on page 77.

Lemma 5.11. *Let B be a deterministic finite automaton with input alphabet Σ and state set Q, and let S and T be two states of $A' = \mathcal{P}(B^R)$. Further, let $m = \max\{\text{level}_{A'}(S), \text{level}_{A'}(T)\}$ and $\ell \geq 0$ be an integer. Then $S \approx_\ell T$ if and only if $S \triangle T \subseteq Q(\Sigma^{> \ell - m})$, that is, if and only if $\text{level}_B(p) > \ell - m$, for all $p \in S \triangle T$.* $\square$

We know from the works of Câmpeanu et al. [17] and Körner [76] that a minimal cover-automaton can be obtained from a given DFA A by merging state p to state q, whenever p and q are similar and $\text{level}_A(p) \geq \text{level}_A(q)$. This leads to the following algorithm for constructing a cover-minimal, or $\Sigma^{>\ell}$-minimal, deterministic finite automaton—see Algorithm 2. Given a DFA A and a cover length ℓ, construct the DFA $B = \mathcal{P}(A^R)$ and compute the sets $Q_B(\Sigma^{>i})$, for $0 \leq i \leq \ell$, by computing the levels of the states in B. Then construct the DFA $A' = \mathcal{P}(B^R)$ and mark each state S in A' with $\text{level}_{A'}(S)$. Finally merge state S to state T, whenever $m = \text{level}_{A'}(S)$ and $S \triangle T \subseteq Q_B(\Sigma^{>\ell - m})$.

Algorithm 2 Brzozowski-like algorithm for finding minimal cover automaton

Require: a DFA or NFA $A = (Q, \Sigma, \delta, q_0, F)$, integer $\ell \geq 0$
1: construct $B = \mathcal{P}(A^R)$
2: **for all** $i = 0, 1, \ldots, \ell$ **do**
3:     compute $Q_B(\Sigma^{>i})$
4: construct $A' = \mathcal{P}(B^R)$
5: **for all** states S of A' (in BFS visit order) **do**
6:     let $m = \text{level}_{A'}(S)$
7:     **find** $T \neq S$ such that $S \triangle T \subseteq Q_B(\Sigma^{>\ell - m})$
8:     **if** T exists **then**
9:         merge S to T
10: **return** A'

Theorem 5.12 (Brzozowski-Like Cover-Minimization). *Given a (deterministic or nondeterministic) finite automaton A, and an integer $\ell \geq 0$, Algorithm 2 computes a cover-minimal deterministic finite automaton A' for the language $L(A) \cap \Sigma^{\leq \ell}$.*

Proof. The DFA A' in line 4 is a minimal DFA, thus it has no unreachable states and no pair of equivalent states. After the `for` loop that starts in line 5 is finished, there is no pair of similar states in A' neither, so by the characterization of cover-minimal DFAs of Körner [76], automaton A' is a cover-minimal DFA for $L(A) \cap \Sigma^{\leq \ell}$. □

The levels of the states of the automata $B = \mathcal{P}(A^R)$ and $A' = \mathcal{P}(B^R)$ can easily be computed on-the-fly, if both power-set constructions are implemented as a breadth-first search. Further, this also allows us to perform the state merging already during the second power-set construction: as soon as we discover a state S on level m that is $\Sigma^{>\ell-m}$-equivalent to some previously discovered state T, we can merge S to T. Note that this is different from the hyper-minimization algorithm in Section 5.3, where we do not start the merging before the whole state graph of $A' = \mathcal{P}(B^R)$ is discovered.

The search in line 7 for a previously discovered state T, to which a newly discovered state S in $A' = \mathcal{P}(B^R)$ could be merged, can be done by cycling through all states of A'. But again, this can be done more cleverly by using hash tables, similar as in Section 5.3. This time we need several hash tables $H_0, H_1, \ldots, H_\ell$, since we also work with several relations, depending on the level of the currently processed state. An entry $H_i[K] = T$ means that $T \setminus Q_B(\Sigma^{>i}) = K$. So if we have some other state S, with $S \setminus Q_B(\Sigma^{>i}) = K$, then we know that $S \sim_{E'} T$, for $E' = \Sigma^{>i}$. We initialize and use these hash tables as follows. Whenever a state S with $m = \text{level}_{A'}(S)$ is discovered, we compute the set of *critical states* $K = S \setminus Q_B(\Sigma^{>\ell-m})$ and check, whether there exists an entry $H_{\ell-m}[K]$ in the hash table $H_{\ell-m}$ for key K. If $H_{\ell-m}[K] = T$, then we merge S to T—we know that $\text{level}_{A'}(S) \geq \text{level}_{A'}(T)$, because we discovered T before S, and we also know that $T \setminus Q(\Sigma^{>\ell-m}) = K$, so $S \triangle T \subseteq Q_B(\Sigma^{>\ell-m})$. If $H_{\ell-m}[K]$ is not defined yet, then we assign $H_i[S \setminus Q(\Sigma^{>i})] = S$, for $0 \leq i \leq \ell - m$. Compared to Brzozowski's minimization algorithm, the only additional resources we need are the $\ell + 1$ hash tables $H_0, H_1, \ldots, H_\ell$.

We illustrate the algorithm with the following example.

Example 5.13. Consider again the DFAs A, $B = \mathcal{P}(A^R)$, and $A' = \mathcal{P}(B^R)$ from Example 5.2—the DFA A' is also depicted on the left-hand side of Figure 5.7. Assume we want to find a minimal cover automaton for A with cover

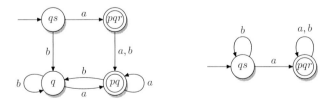

Figure 5.7: Left: the minimal DFA $A' = \mathcal{P}(B^R)$ for the language $L(A)$. Right: the minimal cover DFA A_{cover} for the language $L(A) \cap \Sigma^{\leq 2}$.

length $\ell = 2$, that is, a minimal cover DFA for $L(A) \cap \Sigma^{\leq 2}$. The sets $Q_B(\Sigma^{>i})$, for $0 \leq i \leq \ell$ are

$$Q_B(\Sigma^{>0}) = \{q, r, s, t\}, \qquad Q_B(\Sigma^{>1}) = \{s, t\}, \qquad \text{and} \qquad Q_B(\Sigma^{>2}) = \emptyset.$$

In the for all loop we first process the initial state $S = \{q, s\}$ on level 0. The set of *critical states* is $K - S \setminus Q_B(\Sigma^{>2}) = \{q, s\}$. Since all hash tables are still empty, we do not find an appropriate state T, and so we set $H_i[S \backslash Q_B(\Sigma^{>i})] = S$, for $0 \leq i \leq 2$, which gives

$$H_0[\emptyset] = \{q, s\}, \qquad H_1[\{q\}] = \{q, s\}, \qquad \text{and} \qquad H_2[\{q, s\}] = \{q, s\}.$$

Now assume that from the two states in A' on level 1 we first process the state $S = \{p, q, r\}$. The *critical states* are $K = S \setminus Q_B(\Sigma^{>1}) = \{p, q, r\}$. Since $H_1[K]$ is not defined, we do not find an appropriate state T and add the entries $H_i[S \setminus Q_B(\Sigma^{>i})] = S$, for $0 \leq i \leq 1$ to the hash tables:

$$H_0[\{p\}] = \{p, q, r\}, \qquad \text{and} \qquad H_1[\{p, q, r\}] = \{p, q, r\}.$$

Next we need to process state $S = \{q\}$ on level 1. Here the set of *critical states* is $K = S \setminus Q_B(\Sigma^{>1}) = \{q\}$. We find a matching entry in the hash table $H_{\ell-m} = H_1$, namely $T = H_1[\{q\}] = \{q, s\}$. This means, that we now merge state $S = \{q\}$ to state $T = \{q, s\}$. Since after that, state $\{q\}$ no longer exists in A', we do not have to explore the state graph "behind" $\{q\}$ any further. But we still process the remaining state $\{p, q\}$, because it gets discovered from state $\{p, q, r\}$. So now we have $S = \{p, q\}$ on level 2, with the set of *critical states* $K = S \setminus Q_B(\Sigma^{>0}) = \{p\}$. In the corresponding hash table H_0 we now find the entry $T = H_0[\{p\}] = \{p, q, r\}$, so we merge state $S = \{p, q\}$ to state $T = \{p, q, r\}$. Since we processed all states in A', the algorithm terminates and we obtain the minimal cover DFA for $L(A) \cap \Sigma^{\leq 2}$, which is depicted on the right-hand side of Figure 5.7 on the preceding page. $\diamond$

5.5 k-Minimization

Closely related to the concept of almost-equivalence and hyper-minimality are the notions of k-equivalence and k-minimality—recall Section 3.3: two languages (or automata) are k-equivalent, if they are E-equivalent, for $E = \Sigma^{\leq k}$. Similar to the previous sections, we present an algorithm for k-minimizing deterministic finite automata that constructs the minimal DFA by using reversal and power-set construction twice, and then merges states to obtain a k-minimal automaton. But first we need some notation as used by Gawrychowski et al. [34, 35]. Let $A = (Q, \Sigma, \delta, q_0, F)$ be a deterministic finite automaton and assume that all states in Q are reachable. For a state $q \in Q$, let

$$\text{in-level}_A(q) = \begin{cases} \max\{ |w| \mid w \in L_q \} & \text{if } |L_q| < \infty, \\ \infty & \text{otherwise,} \end{cases}$$

that is, in-level$_A(q)$ denotes the length of the longest word leading to from q_0 to state q, if such a word exists, otherwise in-level$_A(q) = \infty$. Now, for an integer $k \geq 0$, we say that two states p and q are k-similar, denoted by $p \sim_k q$, if

$$d(p, q) + \min\{\text{in-level}_A(p), \text{in-level}_A(q), k\} \leq k.$$

Here $d(p, q)$ is the distance function, which was defined in Section 3.1. Informally, $d(p, q) - 1$ is the length of the longest word that distinguishes between the states p and q. If p and q are equivalent, then $d(p, q) = 0$, and if they are not even almost-equivalent, then $d(p, q) = \infty$. Observe that k-similarity of states can be expressed in terms of E-equivalence as follows.

Lemma 5.14. *Let p and q be two states of a deterministic finite automaton A over alphabet Σ, let $k \geq 0$ and $m = \min\{\text{in-level}_A(p), \text{in-level}_A(q), k\}$. Then p and q are k-similar if and only if they are E-equivalent, for $E = \Sigma^{<k-m}$.*

Proof. It follows from the definition of the $\sim_k$ relation for states that p and q are k-similar if and only if $d(p, q) - 1 < k - m$. By definition of $d(p, q)$ this is equivalent to $L(p) \triangle L(q) \subseteq \Sigma^{<k-m}$, which proves the lemma. $\square$

Now we can use the Lemma 5.5 again, to characterize k-similar states of the deterministic finite automaton $A' = \mathcal{P}(B^R)$—see also Table 5.1 on page 77.

Lemma 5.15. *Let B be a deterministic finite automaton with input alphabet Σ and state set Q, and let S and T be two states of $A' = \mathcal{P}(B^R)$, with*

$$m = \min\{k, \text{in-level}_{A'}(S), \text{in-level}_{A'}(T)\},$$

for an integer $k \geq 0$. Then $S \sim_k T$ if and only if $S \triangle T \subseteq Q_B(\Sigma^{<k-m})$, that is, if and only if in-level$_B(p) < k - m$, for all $p \in S \triangle T$. $\square$

As discussed by Gawrychowski et al. [35], the k-minimal DFA can be obtained by merging k-similar states, where the state with the lower in-level has to be merged to the state with the higher in-level. This enables us to use the the the following algorithm for k-minimization—see Algorithm 3 on the next page. Given a finite automaton A, and an integer k, we first construct the DFA $B = \mathcal{P}(A^R)$, and compute the sets $Q_B(\Sigma^{<i})$, for $0 \leq i \leq k$. Then we build the DFA $A' = \mathcal{P}(B^R)$, and perform the merging of k-similar states.

Theorem 5.16 (Brzozowski-Like k-Minimization). *Given a (deterministic or nondeterministic) finite automaton A and an integer k, Algorithm 3 computes a k-minimal deterministic finite automaton A' that is k-equivalent to A.*

Proof. To perform the merging of k-similar states, it is sufficient to consider only states S with in-level$_A(S) < k$, since any pair of k-similar states S and T with greater in-levels would be a pair of equivalent states, which cannot exist, since the DFA A' in line 4 is a minimal DFA. Further, if we find in line 7 a state T to which a currently processed state S, with $m = \text{in-level}_{A'}(S)$, will be merged, then we know that $S \sim_k T$ holds, because $S \triangle T \subseteq Q(\Sigma^{<k-m})$ implies

---

Algorithm 3 Brzozowski-like algorithm for k-minimizing finite automata

---

Require: a DFA or NFA $A = (Q, \Sigma, \delta, q_0, F)$, integer $k \geq 0$

 1: construct $B = \mathcal{P}(A^R)$
 2: **for all** $i = 0, 1, \ldots, k$ **do**
 3:    compute $Q_B(\Sigma^{<i})$
 4: construct $A' = \mathcal{P}(B^R)$
 5: **for all** states S of A' with in-level$_{A'}(S) < k$ **do**
 6:    let $m = $ in-level$_{A'}(S)$
 7:    **find** $T \neq S$ such that $S \triangle T \subseteq Q_B(\Sigma^{<k-m})$
 8:    **if** T exists, and $m \leq $ in-level$_{A'}(T)$ **then**
 9:       merge S to T
10: **return** A'

$S \triangle T \subseteq Q(\Sigma^{<k-m'})$, for $m' = \min(k, \text{in-level}_{A'}(S), \text{in-level}_{A'}(T)) \leq m$. After the **for** loop that starts in line 5 is finished, there is no pair of k-similar states in A', so A' is cover-minimal. Since only k-similar states were merged, while taking into account their respective levels, the resulting DFA A' is k-equivalent to A. □

The computation of the in-levels and of the sets $Q_B(\Sigma^{<i})$ can be done by a depth-first search algorithm on the reverse of the underlying state graph. The search for an appropriate state T to which a state S can be merged can be implemented with the help of hash tables, similar as in the case of cover automata.

Part III

Biautomata

Chapter 6

Basic Properties of Biautomata

O UR studies in Part II concentrated on finite automata reading their input once from left to right. In the literature one can find plenty of other variations of finite automata which may read the input in different ways. For example two-way finite automata were already considered by Rabin and Scott [105]. In such automata the reading head is not restricted to always move forward on the input tape, that is, from left to right, it may also move backward, from right to left. However, the descriptive capacity of these automata is equal to that of ordinary (one-way) finite automata [105]. Another extension of finite automata is the idea of using more than one reading head. The study of such multi-head finite automata was initiated by Arnold L. Rosenberg [106]—finite automata with more than one reading head were also studied by Rabin and Scott [105], but in such a way that every head belongs to a separate input tape. Again, there exist one-way and two-way versions of these automaton types. The descriptive power of multi-head automata quickly outperforms finite automata: already one-way DFAs with two reading heads can accept non-context-free languages, like the language $\{\, wcw \mid w \in \{a, b\}^* \,\}$ over alphabet $\{a, b, c\}$. A survey of recent results on multi-head finite automata is given by Markus Holzer, Martin Kutrib, and Andreas Malcher [60].

Recently biautomata were introduced by Ondřej Klíma and Libor Polák [74], and the remaining part of this work is devoted to a diverse study of these automata. Informally, a biautomaton can be seen as a finite automaton with two reading heads, where the first head reads the input tape from left to right, starting from the leftmost input symbol, and the second had reads the input from right to left, starting from the rightmost symbol. The head that is used for the current transition is chosen nondeterministically. The computation ends when both heads meet at some position of the input word, that is, when the whole word is consumed.

The biautomaton model as it was introduced by Klíma and Polák [74] is a deterministic machine that has some additional requirements to its transition function, which imply the regularity of the accepted language. These requirements, which we call the $\diamond$-property and the F-property, and their effect on the way the biautomaton accepts a word will be studied in Section 6.2. Building on these results, Chapter 7 analyzes the language classes defined by different types of biautomata. It turns out that such automata in general characterize the class of linear context-free languages, but restricted types of biautomata, such as the model described by Klíma and Polák [74], accept regular languages, and can also be used to characterize well-known subregular language classes.

The connection between biautomata satisfying the $\diamond$- and F-properties and other models for describing regular languages, such as finite automata or regular expressions, will be drawn in Chapter 8. There, the cost of transforming the representation of a regular language from different models to biautomata will be studied. Further, we also consider conversions between different types of biautomata.

Chapter 9 investigates minimization algorithms for biautomata. We will see that—just as for ordinary finite automata—besides classical state merging algorithms also a Brzozowski-like algorithm based on reversal of the automaton can be used to minimize deterministic biautomata that have the $\diamond$- and the F-property. We also indicate problems with minimizing biautomata that do not have both these properties.

Finally, Chapter 10 revisits the idea of lossy compression from Part II, transferring the notions of almost-equivalence and E-equivalence, and the corresponding concepts of hyper- and E-minimality to biautomata. As a sharp contrast to finite automata, we will see that the hyper-minimization problem is computationally intractable already for deterministic biautomata with the $\diamond$-property and the F-property.

6.1   The Biautomaton Model

Now we introduce the model of a nondeterministic biautomaton. This model is more general than the biautomaton model which was introduced by Klíma and Polák [74], and it resembles the model of nondeterministic linear automata from Roussanka Loukanova [84]. A *nondeterministic biautomaton* (NBiA) is a sixtuple $A = (Q, \Sigma, \cdot, \circ, I, F)$, where

- Q is a finite set of *states*,

- Σ is an *input alphabet*,

- $\cdot \colon Q \times \Sigma \to 2^Q$ is the *forward transition function*,

- $\circ \colon Q \times \Sigma \to 2^Q$ is the *backward transition function*,

- $I \subseteq Q$ is the set of *initial states*, and

- $F \subseteq Q$ is the set of *final* or *accepting states*.

Instead of the name *nondeterministic biautomaton*, we may at times inter-changeably use the term *biautomaton*. This is not to be confused with the less general biautomaton model from Klíma and Polák [74].

As it is common usage in literature on biautomata, we use the infix nota-tion $q \cdot a$ and $q \circ a$ for applying the transition functions, instead of the prefix notation $\cdot (q, a)$ and $\circ (q, a)$. The domains of the transition functions can be ex-tended to $Q \times \Sigma^*$, yielding the extended transition functions $\cdot^* : Q \times \Sigma^* \to 2^Q$ and $\circ^* : Q \times \Sigma^* \to 2^Q$, which are defined in the following way: for every state $q \in Q$ let $q \cdot^* \lambda = q \circ^* \lambda = \{q\}$, and for all words $v \in \Sigma^*$ and letters $a \in \Sigma$ define

$$q \cdot^* av = \bigcup_{p \in (q \cdot a)} p \cdot^* v, \qquad \text{and} \qquad q \circ^* va = \bigcup_{p \in (q \circ a)} p \circ^* v.$$

The definitions of $\cdot^*$ and $\circ^*$ show the intention behind the names *forward* and *backward* transition function. While $\cdot^*$ consumes the word as usual from left to right, the function $\circ^*$ consumes the word backwards, that is, from right to left. These functions can be further extended to the functions $\cdot^Q : 2^Q \times \Sigma^* \to 2^Q$ and $\circ^Q : 2^Q \times \Sigma^* \to 2^Q$, where for all sets $P \subseteq Q$, and words $w \in \Sigma^*$,

$$P \cdot^Q w = \bigcup_{p \in P} p \cdot^* w, \qquad \text{and} \qquad P \circ^Q w = \bigcup_{p \in P} p \circ^* w.$$

We will simply denote the extended transition functions by $\cdot$ and $\circ$, instead of using their superscripted versions.

We say that the NBiA $A = (Q, \Sigma, \cdot, \circ, I, F)$ *accepts* a word $w \in \Sigma^*$ if there are (possibly empty) words $u_1, v_1, u_2, v_2, \ldots, u_k, v_k \in \Sigma^*$, such that w can be written as

$$w = u_1 u_2 \ldots u_k v_k \ldots v_2 v_1,$$

and for which we have

$$[((\ldots ((((I \cdot u_1) \circ v_1) \cdot u_2) \circ v_2) \ldots) \cdot u_k) \circ v_k] \cap F \neq \emptyset. \tag{6.1}$$

The *language accepted* by the automaton A is $L(A) = \{ w \in \Sigma^* \mid A \text{ accepts } w \}$. Two biautomata A and B are *equivalent*, denoted by $A \equiv B$, if $L(A) = L(B)$. A biautomaton A is *minimal* if there exists no biautomaton B with fewer states than A, such that $A \equiv B$.

Similar as in the case of classical finite automata, we define the notions of left and right languages, reachability, and equivalence of states in biau-tomata. In the following let $A = (Q, \Sigma, \cdot, \circ, I, F)$ be a biautomaton. The *left language* of state $q \in Q$ is the language $L_{A,q}$ accepted by the biautomaton $A_q = (Q, \Sigma, \cdot, \circ, I, \{q\})$, which is obtained from A by making state q its sole final state. The *right language* of state q is the language $L_A(q)$ accepted by the biautomaton $_qA = (Q, \Sigma, \cdot, \circ, \{q\}, F)$, obtained from A by choosing state q as single initial state.[1] When the automaton A is clear from the context, we may write L_q instead of $L_{A,q}$, and $L(q)$ instead of $L_A(q)$. We say that the state $q \in Q$ is *reachable* in A if its left language $L_{A,q}$ is not empty. Finally, two states p and q of A are *equivalent*, denoted by $p \equiv q$, if their right languages are equal, that is, if $L_A(p) = L_A(q)$ holds.

The automaton $A = (Q, \Sigma, \cdot, \circ, I, F)$ is a *deterministic biautomaton* (DBiA) if $|I| = 1$, and $|q \cdot a| = |q \circ a| = 1$ for all states $q \in Q$ and symbols $a \in \Sigma$. In this case we write $q \cdot a = p$ instead of $q \cdot a = \{p\}$, and $q \circ a = p$ instead of $q \circ a = \{p\}$, treating $\cdot$ and $\circ$ to be mappings from $Q \times \Sigma$ to Q. Further, if $I = \{q_0\}$, we denote the deterministic biautomaton A by $A = (Q, \Sigma, \cdot, \circ, q_0, F)$, omitting the set braces for the singleton set I. A deterministic biautomaton is *minimal* if there is no equivalent deterministic biautomaton that has fewer states.

Although the name *deterministic biautomaton* may suggest differently, the behaviour of such a machine may still be nondeterministic in the following sense. Given a word $w \in \Sigma^*$, the biautomaton A may read w in different ways, some of which may lead to an accepting state, and some may not. Hence, in order to accept w, the automaton A may have to *guess* an appropriate fragmentation of the word and a corresponding sequence of movements of its reading heads.

Example 6.1. Consider the biautomaton $A = (Q, \Sigma, \cdot, \circ, I, F)$ with set of states $Q = \{q_0, q_1, q_2, q_3, q_4, q_5\}$, input alphabet $\Sigma = \{a, b\}$, initial state set $I = \{q_0\}$, and final state set $F = \{q_4\}$. The forward transition function $\cdot$ is defined as follows,

$\cdot$	q_0	q_1	q_2	q_3	q_4	q_5
a	$\{q_1\}$	$\{q_1\}$	$\{q_2\}$	$\{q_5\}$	$\{q_5\}$	$\{q_5\}$
b	$\{q_5\}$	$\{q_1\}$	$\{q_2\}$	$\{q_5\}$	$\{q_5\}$	$\{q_5\}$

and the backward transition function $\circ$ is defined in the following way:

$\circ$	q_0	q_1	q_2	q_3	q_4	q_5
a	$\{q_2\}$	$\{q_3\}$	$\{q_5\}$	$\{q_2\}$	$\{q_5\}$	$\{q_5\}$
b	$\{q_2\}$	$\{q_3\}$	$\{q_4\}$	$\{q_2\}$	$\{q_5\}$	$\{q_5\}$

Notice that A is a *deterministic* biautomaton, because it has a single initial state and the images of both transition functions are always singleton sets. Therefore, we will omit the set braces for the rest of this example.

As for finite automata, we use state transition diagrams to depict biautomata. The two transition functions of the biautomaton are distinguished by two types of arcs in the diagram: solid arrows denote transitions of the forward transition function $\cdot$, and dashed arrows denote transitions of the backward transition function $\circ$. The DBiA A is depicted in Figure 6.1.

---

[1]Note that the words in the right language $L_A(q)$ of a state q are not necessarily suffixes, that is, "right ends" of words from the language $L(A)$, just as the words in the left language $L_{A,q}$ need not be prefixes of words from $L(A)$. One can see that the words in $L_A(q)$ are rather infixes of words from $L(A)$, while the words from $L_{A,q}$ "parenthesize" those infixes. With this point of view the language $L_A(q)$ could also be called the *inner language* of state q, and $L_{A,q}$ could be called the *outer language* of q. Nevertheless, we want to stick to the names left and right language, instead of introducing new names for these languages.

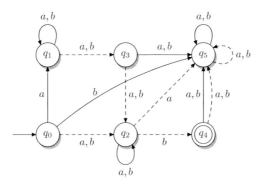

Figure 6.1: The deterministic biautomaton A from Example 6.1 accepting the language $L(A) = a(a + b)^*b(a + b)^2 + (a + b)^*b(a + b)$.

By examining all possible paths from the initial state q_0 to the final state q_4, one can see that the language $L(A)$ accepted by the DBiA A is described by the regular expression

$$a(a + b)^*b(a + b)^2 + (a + b)^*b(a + b).$$

Consider for example the following computation of A on the word $w = abbaa$:

$$
\begin{aligned}
(q_0 \cdot ab) \circ baa &= ((((q_0 \cdot a) \cdot b) \circ a) \circ a) \circ b \\
&= (((q_1 \cdot b) \circ a) \circ a) \circ b \\
&= ((q_1 \circ a) \circ a) \circ b \\
&= (q_3 \circ a) \circ b \\
&= q_2 \circ b \\
&= q_4.
\end{aligned}
$$

Since q_4 is an accepting state, we have

$$[(I \cdot ab) \circ baa] \cap F \neq \emptyset,$$

so the word $w = abbaa$ is accepted by A and hence, belongs to the language $L(A)$. Notice that there is another, different accepting computation of A:

$$[(((q_0 \cdot a) \circ aa) \cdot b) \circ b] \cap F \neq \emptyset.$$

However, the deterministic biautomaton A can also go through a non-accepting computation on w. For example, if A chooses to start reading w by a backward transition, that is, it first reads the a symbol on the right end of the word, then it enters state $q_0 \circ a = q_2$, and the remaining input is the word $abba$. From state q_2 the DBiA A can either read symbols from the beginning of the word, using forward transitions leading from state q_2 to itself, or A can read the symbol a at the end of the remaining input, using a backward transition leading from state q_2 to the non-accepting sink state q_5. $\diamond$

6.2   Structural Properties for Biautomata

Next we define structural restrictions for biautomata and study their effect on the behaviour of the automaton. The first two properties will reduce the influence of the before mentioned nondeterminism of the head movement. These two restrictions were already used by Klíma and Polák [74] as part of the definition of a biautomaton. Let $A = (Q, \Sigma, \cdot, \circ, I, F)$ be a (nondeterministic) biautomaton. The automaton A has the *diamond property*, for short $\diamond$-*property*, if

$$(q \cdot a) \circ b = (q \circ b) \cdot a,$$

holds for every state $q \in Q$ and all symbols $a, b \in \Sigma$. The biautomaton A has the *dual acceptance property*, for short F-*property*, if

$$(q \cdot a) \cap F \neq \emptyset \quad \Longleftrightarrow \quad (q \circ a) \cap F \neq \emptyset,$$

for every state $q \in Q$ and symbol $a \in \Sigma$. The type of biautomaton defined by Klíma and Polák [74] is exactly our definition of a *deterministic* biautomaton that has both the $\diamond$-property and the F-property. Finally, we define another restriction for biautomata, which is similar to the F-property. The biautomaton A has the *equal initial fan-out property*, for short I-*property*, if

$$I \cdot a = I \circ a,$$

for every symbol $a \in \Sigma$. The motivation for this property will become clear in Section 7.3.

We indicate the fact that an NBiA or DBiA has one of the above mentioned properties by using an appropriate prefix. For example a $\diamond$-NBiA is a nondeterministic biautomaton that has the $\diamond$-property and a $\diamond$-F-DBiA is a deterministic biautomaton that has both the $\diamond$- and the F-property.

Example 6.2. Consider again the DBiA A from Example 6.1. This automaton does *not* have any of the three properties defined above. First, the $\diamond$-property is not satisfied, because

$$(q_0 \cdot a) \circ b = q_3 \neq q_2 = (q_0 \circ b) \cdot a.$$

Further, the F property is not satisfied by A in state q_2, because $q_2 \cdot b = q_2 \notin F$ but $q_2 \circ b = q_4 \in F$. To be accurate, we would have use the following notation:

$$(q_2 \cdot b) \cap F = \{q_2\} \cap \{q_4\} = \emptyset \quad \text{but} \quad (q_2 \circ b) \cap F = \{q_4\} \cap \{q_4\} = \{q_4\} \neq \emptyset.$$

Finally, notice that $q_0 \cdot a = q_1 \neq q_2 = q_0 \circ a$, so the biautomaton A neither has the I-property. $\diamond$

The next example presents a nondeterministic biautomaton that *does* satisfy the $\diamond$-property and the F-property.

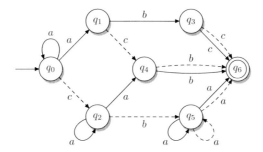

Figure 6.2: The nondeterministic biautomaton A from Example 6.3, that has both the $\diamond$- and the F-property, but not the I-property. The biautomaton accepts the language $L(A) = a^*abc$.

Example 6.3. Let us consider the biautomaton $A = (Q, \Sigma, \cdot, \circ, I, F)$ with state set $Q = \{q_0, q_1, \ldots, q_6\}$, initial state set $I = \{q_0\}$, final state set $F = \{q_6\}$, and the transition functions $\cdot$ and $\circ$ of which are depicted in Figure 6.2. One can check that A has the $\diamond$-property: for all inputs $d, e \in \{a, b, c\}$ and states $q \in Q$ we have $(q \cdot d) \circ e = (q \circ e) \cdot d$. For example, we have

$$(q_0 \cdot a) \circ c = \{q_0, q_1\} \circ c = \{q_2, q_4\} \quad \text{and} \quad (q_0 \circ c) \cdot a = \{q_2\} \cdot a = \{q_2, q_4\}.$$

Further, A has the F-property, that is, for all states $q \in Q$, and symbols $d \in \Sigma$ we have

$$(q \cdot d) \cap F \neq \emptyset \quad \Longleftrightarrow \quad (q \circ d) \cap F \neq \emptyset.$$

For example, both sets $q_1 \cdot b = \{q_3\}$ and $q_1 \circ b = \emptyset$ have an empty intersection with $F = \{q_6\}$, and the sets $q_5 \cdot a = \{q_5, q_6\}$ and $q_5 \circ a = \{q_5, q_6\}$ both contain the accepting state q_6. If we removed the backward transition loop on letter a in state q_5, that is, if $q_5 \circ a = \{q_6\}$, instead of $q_5 \circ a = \{q_5, q_6\}$, then A would not have the $\diamond$-property anymore, because then $(q_5 \cdot a) \circ a \neq (q_5 \circ a) \cdot a$, but it would still have the F-property. One readily sees that A does not fulfill the I-property.

The language accepted by A is $L(A) = a^*abc$. The inclusion $L(A) \supseteq a^*abc$ can easily be seen by considering such paths from the initial state q_0 to the final state q_6, that consist only of forward transitions. We will see later in Corollary 6.11 that it is indeed sufficient to consider only forward transitions to determine the language of a biautomaton that satisfies both the $\diamond$-property and the F-property. $\diamond$

We already mentioned that besides the traditional nondeterministic choice of the next state, the biautomaton also nondeterministically chooses which of its two input heads is moved, in other words, which of the two transition functions is applied in the current step. In the following we show how the influence of the nondeterministic head movement is reduced when the biautomaton obeys the $\diamond$-property, or both the $\diamond$- and the F-property. The proofs of the upcoming results on nondeterministic biautomata are similar to the proofs by Klíma and Polák [74] for the corresponding results on the type of biautomaton defined therein. The next lemma shows that the $\diamond$-property can be extended to words and sets of states.

Lemma 6.4. *Let $A = (Q, \Sigma, \cdot, \circ, I, F)$ be a biautomaton that has the $\diamond$-property. Then $(P \cdot u) \circ v = (P \circ v) \cdot u$, for all sets of states $P \subseteq Q$ and all words $u, v \in \Sigma^*$.*

Proof. By definition of the $\diamond$-property we have $(q \cdot a) \circ b = (q \circ b) \cdot a$ for all states $q \in Q$ and symbols $a, b \in \Sigma$. Then for all sets of states $P \subseteq Q$ and input symbols $a, b \in \Sigma$ we have

$$\left(\bigcup_{p \in P} p \cdot a \right) \circ b = \bigcup_{p \in P} ((p \cdot a) \circ b) = \bigcup_{p \in P} ((p \circ b) \cdot a) = \left(\bigcup_{p \in P} p \circ b \right) \cdot a.$$

This shows that $(P \cdot a) \circ b = (P \circ b) \cdot a$, for all $P \subseteq Q$ and $a, b \in \Sigma$. It remains to show that this equation also holds for words from Σ^* instead of single symbols.

First we prove that $(S \cdot a) \circ v = (S \circ v) \cdot a$ holds for every subset $S \subseteq Q$, word $v \in \Sigma^*$, and symbol $a \in \Sigma$, by induction on the length of v. For the induction basis $|v| = 0$, that is, for $v = \lambda$, we have $(S \cdot a) \circ v = S \cdot a = (S \circ v) \cdot a$. Now let $|v| \geq 1$, so the word v can be written as $v = v'b$, for some symbol $b \in \Sigma$ and a word $v' \in \Sigma^*$. Then

$$(S \cdot a) \circ v = (S \cdot a) \circ v'b = ((S \cdot a) \circ b) \circ v' = ((S \circ b) \cdot a) \circ v'.$$

Now we can use the inductive assumption on v' and obtain

$$((S \circ b) \cdot a) \circ v' = ((S \circ b) \circ v') \cdot a = (S \circ v'b) \cdot a = (S \circ v) \cdot a.$$

Thus, we have shown $(S \cdot a) \circ v = (S \circ v) \cdot a$, for every $S \subseteq Q$, $a \in \Sigma$, and $v \in \Sigma^*$.

Now we can prove the statement of the lemma by performing induction on the length of the word $u \in \Sigma^*$. For the induction basis with $|u| = 0$, that is, $u = \lambda$, we have $(S \cdot u) \circ v = S \circ v = (S \circ v) \cdot u$. Now let $|u| \geq 1$. Assume that $u = au'$, for some symbol $a \in \Sigma$ and a word $u' \in \Sigma^*$. Then

$$(S \cdot u) \circ v = (S \cdot au') \circ v = ((S \cdot a) \cdot u') \circ v,$$

and by using first the inductive assumption on u', and then the statement from above, we obtain

$$((S \cdot a) \cdot u') \circ v = ((S \cdot a) \circ v) \cdot u' = ((S \circ v) \cdot a) \cdot u' = (S \circ v) \cdot au' = (S \circ v) \cdot u,$$

which proves the statement of the lemma. $\square$

By iteratively using Lemma 6.4, the acceptance condition in Equation (6.1) on page 93 can be simplified as follows.

Lemma 6.5. *Let $A = (Q, \Sigma, \cdot, \circ, I, F)$ be a biautomaton that satisfies the $\diamond$-property. Then for all words $u_1, v_1, u_2, v_2, \ldots, u_k, v_k \in \Sigma^*$ and all subsets of states $P \subseteq Q$ we have*

$$((\ldots ((((P \cdot u_1) \circ v_1) \cdot u_2) \circ v_2) \ldots) \cdot u_k) \circ v_k = (P \cdot u_1 u_2 \ldots u_k) \circ v_k \ldots v_2 v_1.$$

Proof. By iteratively applying Lemma 6.4, the left-hand side of the equation can be transformed into the right-hand side. □

Lemma 6.4 implies that the order in which the reading heads of a $\diamond$-NBiA move is not important. Therefore, we immediately get the following simple acceptance conditions for $\diamond$-NBiAs.

Corollary 6.6. *Let $A = (Q, \Sigma, \cdot, \circ, I, F)$ be a biautomaton that satisfies the $\diamond$-property. Then A accepts a word $w \in \Sigma^*$ if and only if there are words $u, v \in \Sigma^*$, with $w = uv$, such that*

$$[(I \cdot u) \circ v] \cap F \neq \emptyset.$$

Therefore, the language accepted by A is

$$L(A) = \{\, w \in \Sigma^* \mid \exists u, v \in \Sigma^* : w = uv \text{ and } [(I \cdot u) \circ v] \cap F \neq \emptyset \,\}. \qquad □$$

Similarly to the $\diamond$-property, also the F-property can be extended to sets of states, as the following result shows.

Lemma 6.7. *Let $A = (Q, \Sigma, \cdot, \circ, I, F)$ be a biautomaton that has the F-property. Then $(P \cdot a) \cap F \neq \emptyset$ if and only if $(P \circ a) \cap F \neq \emptyset$, for all state sets $P \subseteq Q$ and all symbols $a \in \Sigma$.*

Proof. For all sets $P \subseteq Q$ and symbols $a \in \Sigma$ we have

$$(P \cdot a) \cap F = \Big(\bigcup_{p \in P} p \cdot a \Big) \cap F = \Big(\bigcup_{p \in P} (p \cdot a) \cap F \Big),$$

and

$$(P \circ a) \cap F = \Big(\bigcup_{p \in P} p \circ a \Big) \cap F = \Big(\bigcup_{p \in P} (p \circ a) \cap F \Big).$$

If A has the F-property, then $(p \cdot a) \cap F \neq \emptyset$ if and only if $(p \circ a) \cap F \neq \emptyset$, for all states $p \in P$ and symbols $a \in \Sigma$. The statement of the lemma follows. □

Note that the F-property does *not* extend to words instead of symbols, as it is the case for the $\diamond$-property. This is demonstrated in the following example.

Example 6.8. Consider the NBiA A over alphabet $\Sigma = \{a\}$, with state set $Q = \{q_0, q_1\}$, initial state q_0, final state set $F = \{q_1\}$ and the transitions:

$$q_0 \cdot a = \{q_0, q_1\}, \qquad q_0 \circ a = \{q_1\}, \qquad q_1 \cdot a = q_1 \circ a = \emptyset.$$

One can see that biautomaton A has the F-property, but $(q_0 \cdot aa) \cap F = \{q_1\}$ and $(q_0 \circ aa) \cap F = \emptyset$. So the F-property does not extend to words. ◇

We have seen in Corollary 6.6 that in order to accept a word $w \in \Sigma^*$, a $\diamond$-NBiA only needs to guess the position in $w = uv$, that splits the word in a first part $u \in \Sigma^*$, which should be read with forward transitions, and a second part $v \in \Sigma^*$, that should be read with backward transitions. If the biautomaton satisfies the F-property in addition to the $\diamond$-property, then also this nondeterministic guess is not important anymore as the upcoming results show.

Lemma 6.9. *Let $A = (Q, \Sigma, \cdot, \circ, I, F)$ be a biautomaton with the $\diamond$-property and the F-property. Then for every set $P \subseteq Q$ and all words $u, v, w \in \Sigma^*$,*

$$[(P \cdot uv) \circ w] \cap F \neq \emptyset \quad \Longleftrightarrow \quad [(P \cdot u) \circ vw] \cap F \neq \emptyset.$$

Proof. We prove the statement by induction on the length of v. Note that the statement holds for $|v| = 0$, because for $v = \lambda$ we have $(P \cdot uv) \circ w = (P \cdot u) \circ vw$. Now let $|v| \geq 1$. In this case the word v can be written as $v = v'a$, for some word $v' \in \Sigma^*$ and a symbol $a \in \Sigma$. We can use Lemma 6.4 to see that

$$(P \cdot uv) \circ w = ((P \cdot uv') \cdot a) \circ w = ((P \cdot uv') \circ w) \cdot a.$$

Hence, $[(P \cdot uv) \circ w] \cap F \neq \emptyset$ if and only if $[((P \cdot uv') \circ w) \cdot a] \cap F \neq \emptyset$. Since A has the F-property, the latter holds if and only if $[((P \cdot uv') \circ w) \circ a] \cap F \neq \emptyset$. Observe that $[((P \cdot uv') \circ w) \circ a]$ is equal to $[(P \cdot uv') \circ aw]$. Now we use the inductive assumption on v', and see that $[(P \cdot uv') \circ aw] \cap F \neq \emptyset$ holds if and only if $[(P \cdot u) \circ v'aw] \cap F \neq \emptyset$, which in turn is equivalent to $[(P \cdot u) \circ vw] \cap F \neq \emptyset$. This concludes the proof. $\qquad \square$

Iterated application of Lemma 6.9 implies the following result.

Lemma 6.10. *Let $A = (Q, \Sigma, \cdot, \circ, I, F)$ be a biautomaton that satisfies the $\diamond$-property and the F-property. Then for all words $u_1, v_1, u_2, v_2, \ldots, u_k, v_k \in \Sigma^*$ and all subsets of states $P \subseteq Q$ we have*

$$[((\ldots ((((P \cdot u_1) \circ v_1) \cdot u_2) \circ v_2) \ldots) \cdot u_k) \circ v_k] \cap F \neq \emptyset$$

if and only if

$$(P \cdot u_1 u_2 \ldots u_k v_k \ldots v_2 v_1) \cap F \neq \emptyset.$$

Proof. Let $u = u_1 u_2 \ldots u_k$ and $v = v_k \ldots v_2 v_1$. Lemma 6.5 implies

$$((\ldots ((((P \cdot u_1) \circ v_1) \cdot u_2) \circ v_2) \ldots) \cdot u_k) \circ v_k = (P \cdot u) \circ v.$$

Now we can use Lemma 6.9 to see that $[(P \cdot u) \circ v\lambda] \cap F \neq \emptyset$ holds if and only if $[(P \cdot uv) \circ \lambda] \cap F \neq \emptyset$. Therefore

$$[((\ldots ((((P \cdot u_1) \circ v_1) \cdot u_2) \circ v_2) \ldots) \cdot u_k) \circ v_k] \cap F \neq \emptyset$$

holds if and only if $(P \cdot u_1 u_2 \ldots u_k v_k \ldots v_2 v_1) \cap F \neq \emptyset$. $\qquad \square$

Notice that Lemma 6.10 implies that for acceptance or non-acceptance of an input word, it does not matter at all, in which order the reading heads are moved and at which position of the input word the two reading heads meet. Either all possible computations of a $\diamond$-F-NBiA on an input word w are accepting, or all are non-accepting. Now the acceptance condition from Equation (6.1) becomes even simpler, if the biautomaton has both the $\diamond$-property and the F-property.

Corollary 6.11. *Let $A = (Q, \Sigma, \cdot, \circ, I, F)$ be a biautomaton that satisfies the $\diamond$-property and the F-property. Then A accepts a word $w \in \Sigma^*$ if and only if*

$$(I \cdot w) \cap F \neq \emptyset.$$

Therefore, the language accepted by A is $L(A) = \{\, w \in \Sigma^* \mid (I \cdot w) \cap F \neq \emptyset \,\}$.   □

If we apply Corollary 6.11 to the $\diamond$-F-NBiA A from Example 6.3, which is depicted in Figure 6.2 on page 97, we readily see that $L(A) = a^* abc$.

It remains to discuss the I-property. If we consider biautomata with the $\diamond$-property, and the I-property, then switching from $\cdot$ to $\circ$ when reading a subword induces a circular shift on the word, which can be seen in the following lemma.

Lemma 6.12. *Let $A = (Q, \Sigma, \cdot, \circ, I, F)$ be a biautomaton that satisfies the $\diamond$-property and the I-property. Then, for all words $u, v, w \in \Sigma^*$ we have*

$$(I \cdot uv) \circ w = (I \cdot v) \circ wu.$$

Proof. We use induction on the length of the word u. For $|u| = 0$, that is, in case $u = \lambda$, we have $(I \cdot uv) \circ w = (I \cdot v) \circ w = (I \cdot v) \circ wu$. Now let $|u| \geq 1$ and assume that $u = au'$, for some symbol $a \in \Sigma$ and word $u' \in \Sigma^*$. By Lemma 6.4, together with the I-property, we obtain

$$(I \cdot uv) \circ w = ((I \cdot a) \cdot u'v) \circ w = ((I \circ a) \cdot u'v) \circ w = ((I \cdot u'v) \circ a) \circ w = (I \cdot u'v) \circ wa.$$

Then the proof can be concluded by using the inductive assumption on the word u' to obtain $(I \cdot u'v) \circ wa = (I \cdot v) \circ wau' = (I \cdot v) \circ wu$.   □

6.2.1   Biautomata with $\diamond$- and F-property

Recall that the biautomaton model as it was introduced by Klíma and Polák [74] is exactly the model of a deterministic biautomaton that has both the $\diamond$-property and the F-property. This model is deterministic in its state transitions, and the power of the nondeterministic head movement is annihilated by the two structural properties. It turns out that $\diamond$-F-DBiAs and classical DFAs share a lot of similarities. This is witnessed for example by Corollary 6.11, which shows that every biautomaton with $\diamond$- and F-property *contains* an equivalent finite automaton in the following sense.

Fact 6.13. *Let $A = (Q, \Sigma, \cdot, \circ, I, F)$ be a nondeterministic biautomaton that satisfies the $\diamond$-property and the F-property. Then $A' = (Q, \Sigma, \delta, I, F)$ with $\delta(q, a) = q \cdot a$, for all states $q \in Q$ and symbols $a \in \Sigma$, is a nondeterministic finite automaton (with nondeterministic initial state), with $L(A') = L(A)$. Moreover, if A is deterministic, then A' is deterministic as well.*

When we use the backward transition function $\circ$ of the biautomaton A to define the transition function δ of the finite automaton, then we obtain a finite automaton that accepts the reversal of the language accepted by the biautomaton:

Fact 6.14. *Let* $A = (Q, \Sigma, \cdot, \circ, I, F)$ *be a nondeterministic biautomaton that satisfies the $\diamond$-property and the F-property. Then* $A' = (Q, \Sigma, \delta, I, F)$ *with* $\delta(q, a) = q \circ a$, *for all states* $q \in Q$ *and symbols* $a \in \Sigma$, *is a nondeterministic finite automaton (with nondeterministic initial state), with* $L(A') = L(A)^R$. *Moreover, if A is deterministic, then A' is deterministic as well.*

Therefore, every language accepted by some deterministic or nondeterministic biautomaton with the $\diamond$- and the F-property is a regular language. On the other hand, different ways of constructing a $\diamond$-F-DBiA from a given DFA were presented by Klíma and Polák [74], so the class of languages accepted by deterministic biautomata coincides with the class of regular languages. Further relations between different biautomata models and known language classes will be studied in Chapter 7.

In the following, we present a useful concept that shows further interesting similarities between $\diamond$-F-DBiAs and deterministic finite automata, namely the construction of the canonical biautomaton [74]. This construction is very similar to the description of the minimal deterministic finite automaton of a regular language by derivatives of the language, as presented by Brzozowski [13]. Before we describe the canonical biautomaton, recall the following definition from Section 2.1. For a language $L \subseteq \Sigma^*$ and words $u, v \in \Sigma^*$, let

$$u^{-1}Lv^{-1} = \{\, w \in \Sigma^* \mid uwv \in L \,\}.$$

Further remember that by Lemma 2.2, the number of both-sided derivatives of a regular language is finite.

Now the *canonical biautomaton* [74] of a regular language $L \subseteq \Sigma^*$ is the deterministic biautomaton $A_L = (Q_L, \Sigma, \cdot, \circ, q_0^{(L)}, F_L)$, where the set of states is

$$Q_L = \{\, u^{-1}Lv^{-1} \mid u, v \in \Sigma^* \,\},$$

the single initial state is $q_0^{(L)} = L$, the set of final states is

$$F_L = \{\, L' \in Q \mid \lambda \in L' \,\},$$

and the transition functions $\cdot$ and $\circ$ are defined as follows, for all states $L' \in Q_L$ and input symbols $a \in \Sigma$: $L' \cdot a = a^{-1}L'$, and $L' \circ a = L'a^{-1}$. Notice that this function generalizes to words $w \in \Sigma^*$ as follows:

$$L' \cdot w = w^{-1}L' \qquad \text{and} \qquad L' \circ w = L'w^{-1}.$$

One can see that the automaton A_L is a $\diamond$-F-DBiA: the $\diamond$-property is satisfied, because

$$(L' \cdot a) \circ b = \{\, w \in \Sigma^* \mid awb \in L \,\} = (L' \circ b) \cdot a,$$

for all states L' of the canonical biautomaton, and all inputs $a, b \in \Sigma$. Further, both conditions $\lambda \in a^{-1}L'$ and $\lambda \in L'a^{-1}$ are equivalent, because both hold if and only if $a \in L'$. Therefore, the automaton A_L also has the F-property.

It was shown by Klíma and Polák [74] that the right language of a state L' of the canonical biautomaton A_L is the language L'. For clarity, we recall the argumentation here. Let $q = u^{-1}Lv^{-1} \in Q_L$ be a state of the canonical biautomaton A_L. Then a word $w \in \Sigma^*$ belongs to the right language $L_{A_L}(q)$ of state q if and only if $q \cdot w \in F_L$. Note that the state

$$ q \cdot w = (u^{-1}Lv^{-1}) \cdot w = w^{-1}(u^{-1}Lv^{-1}) $$

is an accepting state of A_L if and only if $\lambda \in w^{-1}(u^{-1}Lv^{-1})$, that is, if and only if $w \in u^{-1}Lv^{-1}$. This shows $L_{A_L}(q) = u^{-1}Lv^{-1}$.

Just like in the case of finite automata, it turns out that the canonical biautomaton of a regular language L is minimal (with respect to the number of states) among all deterministic biautomata accepting language L and satisfying the $\diamond$-property and the F-property [74]. Actual minimization algorithms for deterministic biautomata with $\diamond$- and F-property, which are very similar to known minimization algorithm for DFAs, will be discussed in Sections 9.1 and 9.2. Nevertheless, besides all these similarities between $\diamond$-F-DBiAs and DFAs, a surprising difference between these two models shows up, when considering lossy compression. We will show in Section 10.2 that the hyper-minimization problem for $\diamond$-F-DBiAs is computationally hard, namely NP-complete. This stands in sharp contrast to the NL-completeness of the hyper-minimization problem for deterministic finite automata, which has been shown in Section 4.2.1.

Chapter 7

Language Classes Related to Different Types of Biautomata

I_N this chapter we investigate the classes of languages that are accepted by biautomata with different structural restrictions. We will see in Section 7.1 that biautomata without any restrictions characterize the class of linear context-free languages, and in Section 7.2 we show that the class of regular languages coincides with the family of languages accepted by biautomata with the $\diamond$-property. Further restrictions on the structure of biautomata will lead to characterizations of some well-known subregular language classes, which will be discussed in Section 7.2.1.

The first step in our study of the language families accepted by different types of biautomata is to show that deterministic and nondeterministic machines are equally powerful. This can be shown by a straightforward power-set construction. For a biautomaton $A = (Q, \Sigma, \cdot, \circ, I, F)$, its *power-set automaton* is the deterministic biautomaton $\mathcal{P}(A) = (Q', \Sigma, \cdot', \circ', q_0', F')$, where the state set is $Q' = 2^Q$, the initial state is $q_0' = I$, the set of accepting states is $F' = \{ P \in Q' \mid P \cap F \neq \emptyset \}$, and the forward and backward transition functions are defined by

$$ P \cdot' a = \bigcup_{p \in P} p \cdot a \qquad \text{and} \qquad P \circ' a = \bigcup_{p \in P} p \circ a, $$

for every state $P \in Q'$ and symbol $a \in \Sigma$.

The next lemma shows that the power-set construction works as expected, and that it preserves the structural properties defined in Section 6.2.

Lemma 7.1. *If A is a nondeterministic biautomaton, then $L(A) = L(\mathcal{P}(A))$, that is, $\mathcal{P}(A)$ is equivalent to A. Furthermore, for all $X \in \{\diamond, F, I\}$, if A has the X-property, then the deterministic biautomaton $\mathcal{P}(A)$ has the X-property, too.*

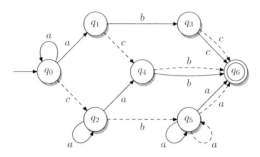

Figure 7.1: The nondeterministic biautomaton A that will be determinized by the power-set construction.

Proof. Consider a nondeterministic biautomaton $A = (Q, \Sigma, \cdot, \circ, I, F)$ and its power-set biautomaton $\mathcal{P}(A) = (Q', \Sigma, \cdot', \circ', q'_0, F')$. For every word $w \in \Sigma^*$, we have $w \in L(A)$ if and only if $w = u_1 u_2 \ldots u_k v_k \ldots v_2 v_1$, for words $u_i, v_i \in \Sigma^*$, with $1 \leq i \leq k$, such that

$$[((\ldots ((((I \cdot u_1) \circ v_1) \cdot u_2) \circ v_2) \ldots) \cdot u_k) \circ v_k] \cap F \neq \emptyset.$$

By definition of the power-set automaton $\mathcal{P}(A)$, this is equivalent to the condition

$$[((\ldots ((((q'_0 \cdot' u_1) \circ' v_1) \cdot' u_2) \circ' v_2) \ldots) \cdot' u_k) \circ' v_k] \in F',$$

which holds if and only if $w \in L(\mathcal{P}(A))$. Hence, $L(A) = L(\mathcal{P}(A))$.

Since the transition functions $\cdot'$ and $\circ'$ of B are just the extensions of the functions $\cdot$ and $\circ$ of A, respectively, to sets of states, the $\diamond$-property, the F-property, and the I-property are preserved by the power-set construction—see Lemmas 6.4 and 6.7. □

We illustrate the power-set construction in the following example.

Example 7.2. Let $A = (Q, \Sigma, \cdot, \circ, I, F)$ be the nondeterministic biautomaton from Example 6.3. For convenience the automaton A is depicted again in Figure 7.1. Recall that A is a $\diamond$-F-NBiA. Its corresponding power-set biautomaton is the deterministic biautomaton $\mathcal{P}(A) = (Q', \Sigma, \cdot', \circ', q'_0, F')$, which is depicted in Figure 7.2. The state labels stand for the corresponding states of $\mathcal{P}(A)$, which are subsets of Q. For example the label $q_0 q_1$ denotes the state $\{q_0, q_1\}$ of $\mathcal{P}(A)$. Because A is a $\diamond$-F-NBiA, also its power-set automaton $\mathcal{P}(A)$ has both the $\diamond$-property and the F-property. $\diamond$

7.1 Linear Context-Free Languages

Our model of nondeterministic biautomata resembles that of linear automata, which were introduced by Loukanova [84] as an automaton model that characterizes the family of linear context-free languages. Therefore, the class of languages

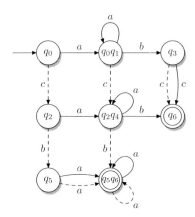

Figure 7.2: The power-set automaton $\mathcal{P}(A)$ of the nondeterministic biautomaton A from Figure 7.1. The non-accepting sink state $\emptyset$, and all transitions leading to it, are not shown. Further, the states which are not reachable from the initial state are also not depicted.

accepted by nondeterministic biautomata coincides with that language family. In this section we want to prove this result directly in terms of biautomata, both for completeness of this work, and because the obtained results will be used in later sections.

First we need some basics on context-free languages, as to be found, for example, in the books of Hopcroft et al. [62] or Michael A. Harrison [45]. A *context-free grammar* is a quadruple $G = (N, T, P, S)$ consisting of

- a finite set N of *nonterminals*,

- a finite set T of *terminals*, which is disjoint from N,

- a finite set $P \subseteq N \times (N \cup T)^*$ of *productions*, and

- the *axiom* or *start symbol* $S \in N$.

We represent the productions of P in the form $X \to \alpha$ instead of (X, α). We call the nonterminal X the *left-hand side* and the word α the *right-hand side* of the production $X \to \alpha$. The grammar $G = (N, T, P, S)$ is a *linear context-free grammar* if $P \subseteq N \times (T^* N T^* \cup T^*)$, in other words, if the right-hand side of any production is either a terminal word, that is, a word over the alphabet T, or it is a word over $(T \cup N)$ that contains exactly one nonterminal.

Let $G = (N, T, P, S)$ be a linear context-free grammar and $\alpha, \beta \in (T \cup N)^*$. We say that β can be *derived in one step* from α by the grammar G, if $\alpha = uXw$, for some nonterminal $X \in N$ and words $u, w \in (T \cup N)^*$, there is a production $X \to v$ in P, and $\beta = uvw$. If β can be derived from α by G in one step, we denote such a *derivation step* by $\alpha \Rightarrow_G \beta$. The reflexive, transitive closure of the relation $\Rightarrow_G$ is denoted by $\Rightarrow_G^*$. If $\alpha \Rightarrow_G^* \beta$, we say that β can be *derived* from α by G (in an arbitrary number of steps). If the grammar G is clear from the context, we omit the index G and write $\Rightarrow$ and $\Rightarrow^*$, instead of $\Rightarrow_G$ and $\Rightarrow_G^*$, respectively. A word $\alpha \in (N \cup T)^*$ that can be derived from the axiom S, that is, with $S \Rightarrow_G^* \alpha$, is called a *sentential form* of G. Due to the structure of productions of a linear context-free grammar, any sentential form that is not a terminal word contains exactly one nonterminal. Now the *language $L(G)$ generated* by grammar G is the set of all sentential forms of G that are terminal words:

$$L(G) = \{\, w \in T^* \mid S \Rightarrow_G^* w \,\}.$$

The class of *linear context-free languages* is the family of languages that are generated by linear context-free grammars, and will be denoted by LIN.

Example 7.3. Consider the linear context-free grammar $G = (N, T, P, S)$ with the set of nonterminals $N = \{S, X, Y\}$, from which S is the axiom, the set of terminals $T = \{a, b, c\}$, and the following productions:

$$P = \{\, S \to aXbb,\ S \to Y,\ X \to aXbb,\ X \to c,\ Y \to aaYb,\ Y \to \lambda \,\}.$$

Any derivation in G has to start with one of the two derivation steps

$$S \Rightarrow_G aXbb, \qquad \text{or} \qquad S \Rightarrow_G Y.$$

From the sentential from $aXbb$, derivations may now continue by using the production $X \to aXbb$, so we obtain derivations

$$aXbb \Rightarrow_G aaXbbbb \Rightarrow_G \cdots \Rightarrow_G a^i X b^{2i},$$

for integers $i \geq 1$. These derivations can terminate by using the production $X \to c$, resulting in the terminal word $a^i c b^{2i}$. So we have $S \Rightarrow_G^* a^i c b^{2i}$, for all $i \geq 1$. From the sentential form Y, the derivation may continue using the production $Y \to aaYb$, which gives us derivations of the form

$$aaYb \Rightarrow_G aaaaYbb \Rightarrow_G \cdots \Rightarrow_G a^{2i} Y b^i,$$

for $i \geq 0$. These derivations can terminate by using the production $Y \to \lambda$. Therefore, $S \Rightarrow_G^* a^{2i} b^i$, for all $i \geq 0$. One can see that there are no other derivations possible in the grammar G, so the language generated by this grammar is the linear context-free language $L(G) = \{\, a^i c b^{2i} \mid i \geq 1 \,\} \cup \{\, a^{2i} b^i \mid i \geq 0 \,\}$. $\diamond$

A linear context-free grammar $G = (N, T, P, S)$ is in *normal form* if any production in P has one of the forms $X \to aY$, $X \to Ya$, or $X \to a$, for nonterminals $X, Y \in N$ and a terminal $a \in T$. A possible exception to this rule is the production $S \to \lambda$, which is allowed if the nonterminal S does not appear on the right-hand side of any production in P. It is known that any linear context-free grammar G can be effectively transformed into a linear context-free grammar G' in normal form, such that $L(G) = L(G')$ [45]. This can be achieved by replacing every production of the form $X \to uYv$, with $X, Y \in N$ and $u, v \in T^*$ by a sequence of productions that generate the terminal words u and v symbol by symbol, as illustrated in the upcoming example. Further, chain productions of the form $X \to Y$, for $X, Y \in N$, and productions $Z \to \lambda$, for $Z \in V \setminus \{S\}$ can be eliminated—the rule $S \to \lambda$ possibly has to be included to allow generation of the empty word.

Example 7.4. Recall the grammar $G = (N, T, P, S)$ from Example 7.3, with the set of productions

$$P = \{ S \to aXbb, \; S \to Y, \; X \to aXbb, \; X \to c, \; Y \to aaYb, \; Y \to \lambda \}.$$

This grammar is *not* in normal form, because, except for $X \to c$, none of the productions has one of the allowed forms. A linear context-free grammar in normal form, that generates the same language, is $G' = (N', T, S, P')$, with the set of nonterminals

$$N' = \{ S, X, Y, [X_{bb}], [X_b], [_{aa}Y], [_aY] \},$$

the axiom S, and the following set of productions:

$$P' = \{ S \to \lambda, \; S \to a[X_{bb}], \; [X_{bb}] \to [X_b]b, \; [X_b] \to Xb, \; X \to a[X_{bb}], \; X \to c \}$$
$$\cup \{ S \to [_{aa}Y]b, \; [_{aa}Y] \to a[_aY], \; [_aY] \to aY, \; Y \to [_{aa}Y]b, \; [_aY] \to a \}.$$

The productions from P' simulate the productions from P. For example the derivation step $X \Rightarrow_G aXbb$ in G is simulated by grammar G' by the sequence of steps $X \Rightarrow_{G'} a[X_{bb}] \Rightarrow_{G'} a[X_b]b \Rightarrow_{G'} aXbb$. Consider the following derivation of the word $w = aacbbbb$ in the grammar G,

$$S \Rightarrow_G aXbb \Rightarrow_G aaXbbbb \Rightarrow_G aacbbbb,$$

and compare this to the corresponding derivation in G':

$$S \Rightarrow_{G'} a[X_{bb}] \Rightarrow_{G'} a[X_b]b \Rightarrow_{G'} aXbb$$
$$\Rightarrow_{G'} aa[X_{bb}]bb \Rightarrow_{G'} aa[X_b]bbb \Rightarrow_{G'} aaXbbbb$$
$$\Rightarrow_{G'} aacbbbb. \qquad \diamond$$

The way a word is produced letter by letter in a derivation by a linear context-free grammar in normal form suggests how to connect such grammars with biautomata: use the duality of producing a letter by the grammar, and consuming an input symbol by the automaton. We first show how to convert a biautomaton A into a linear context-free grammar G in normal form, such that $L(G) = L(A)$.

Theorem 7.5. *Let A be an n-state nondeterministic biautomaton. Then one can effectively construct a linear context-free grammar G in normal form with $n + 1$ nonterminals, such that $L(G) = L(A)$.*

Proof. Let $A = (Q, \Sigma, \cdot, \circ, I, F)$ be a nondeterministic biautomaton. Then we construct a linear context-free grammar $G = (N, T, P, S)$ in normal form with nonterminals $N = \{ X_q \mid q \in Q \} \cup \{S\}$, terminals $T = \Sigma$, and productions

$$P = \{ S \to \lambda \mid I \cap F \neq \emptyset \} \cup P_{\text{fwd}} \cup P_{\text{bwd}},$$

where the set P_{fwd}, corresponding to the forward transitions of A, is defined as

$$P_{\text{fwd}} = \{ S \to aX_q \mid q \in (I \cdot a) \} \cup \{ S \to a \mid (I \cdot a) \cap F \neq \emptyset \}$$
$$\cup \{ X_p \to aX_q \mid q \in (p \cdot a) \} \cup \{ X_p \to a \mid (p \cdot a) \cap F \neq \emptyset \}$$

and the set P_{bwd} of productions corresponding to backward transitions of A is

$$P_{\text{bwd}} = \{ S \to X_q a \mid q \in (I \circ a) \} \cup \{ S \to a \mid (I \circ a) \cap F \neq \emptyset \}$$
$$\cup \{ X_p \to X_q a \mid q \in (p \circ a) \} \cup \{ X_p \to a \mid (p \circ a) \cap F \neq \emptyset \}.$$

To see that $L(G) = L(A)$, consider a word $w \in \Sigma^*$. We show that $w \in L(G)$ if and only if $w \in L(A)$. Notice that for $w = \lambda$, the biautomaton A accepts w if and only if $I \cap F \neq \emptyset$. By construction of the grammar G this holds if and only if $S \to \lambda$ is a production of G. Since G is in normal form, there is no other way of producing the empty word. Therefore, we have $\lambda \in L(A)$ if and only if $\lambda \in L(G)$. Now consider a word $w = a$, consisting of a single symbol $a \in \Sigma$. One can see that $a \in L(G)$ if and only if $S \to a$ is a production of G. By construction this is true if and only if at least one of the sets $I \cdot a$ and $I \circ a$ has a nonempty intersection with F, that is, if and only if $a \in L(A)$.

In order to show that $w \in L(G)$ if and only if $w \in L(A)$, for all words $w \in \Sigma^*$ with $|w| \geq 2$, we use the following claim, which will be proved later.

Claim. Let $k \geq 1$ and $w = u_1 u_2 \ldots u_k v_k \ldots v_2 v_1$, where the word pairs (u_i, v_i) are of the form $(u_i, v_i) \in (\{\lambda\} \times \Sigma) \cup (\Sigma \times \{\lambda\})$, for $1 \leq i \leq k$. Then, for all $p, q \in Q$,

$$q \in [((\ldots ((((p \cdot u_1) \circ v_1) \cdot u_2) \circ v_2) \ldots) \cdot u_k) \circ v_k]$$

holds if and only if

$$X_p \Rightarrow_G^* u_1 u_2 \ldots u_k X_q v_k \ldots v_2 v_1.$$

Assume the claim to be true and let $w \in \Sigma^*$, with $|w| \geq 2$. Because the grammar is in normal form, any derivation of the word w in G consists of $|w| \geq 2$ steps. Therefore we have $w \in L(G)$ if and only if there are derivations

- $S \Rightarrow_G \alpha$, with $\alpha = aX_p$ or $\alpha = X_p a$, for some $p \in Q$ and $a \in \Sigma$,

- $X_p \Rightarrow_G^* uX_q v$, for some $q \in Q$ and $u, v \in \Sigma^*$,

- and $X_q \Rightarrow_G b$, for some $b \in \Sigma$,

such that $w = aubv$, if $\alpha = aX_p$, and $w = ubva$, if $\alpha = X_pa$. The first derivation step $S \Rightarrow_G \alpha$ is possible in G if and only if G has a production $S \to aX_p$ or $S \to X_pa$, depending on whether $\alpha = aX_p$ or $\alpha = X_pa$, respectively. This is equivalent to the fact that in the biautomaton A we have $p \in I \cdot a$ or $p \in I \circ a$, respectively. By using the claim from above, one can see that the derivation $X_p \Rightarrow_G^* uX_qv$ can be carried out in G if and only if the biautomaton A can read the word u with forward transitions and the word v with backward transitions in some order to go from state p to state q. The final derivation step $X_q \Rightarrow_G b$ is possible in G if and only if there is a production $X_q \to b$. By construction, such a production exists in G if and only if state q of the biautomaton A goes to an accepting state on a forward or backward transition on symbol b. Putting all this together, we see that the word w can be generated by the grammar G if and only if it is accepted by the biautomaton A.

It remains to prove the claim, which we do by induction on k. We start with the case $k = 1$. Then (u_1, v_1) is either (a, λ) or (λ, a), for some symbol $a \in \Sigma$. In the former case we have $(p \cdot u_1) \circ v_1 = (p \cdot a) \circ \lambda = (p \cdot a)$. If $q \in (p \cdot a)$, we find the derivation $X_p \Rightarrow_G aX_a$, because P contains the production $X_p \to aX_q$. Conversely, if the production $X_p \to aX_q$ is in P, then it must be $q \in (p \cdot a)$. Therefore, we have $q \in [(p \cdot a) \circ \lambda]$ if and only if $X_p \Rightarrow_G^* aX_q$. In the second case, where $(u_1, v_1) = (\lambda, a)$, we argue in a similar way: we have $(p \cdot u_1) \circ v_1 = (p \circ a)$ and by construction of G' we have $q \in (p \circ a)$ if and only if $X_p \to X_qa$. Altogether we find that $q \in [(p \cdot u_1) \circ v_1]$ if and only if $X_p \Rightarrow_G^* u_1X_qv_1$, for all states p and q of A.

Next, let $k \geq 1$ and we do the inductive step from k to $k + 1$. Consider the computation $[((\ldots((((p \cdot u_1) \circ v_1) \cdot u_2) \circ v_2) \ldots) \cdot u_{k+1}) \circ v_{k+1}]$. Then by the inductive assumption we have

$$q \in [((\ldots((((p \cdot u_1) \circ v_1) \cdot u_2) \circ v_2) \ldots) \cdot u_k) \circ v_k]$$

if and only if

$$X_p \Rightarrow_G^* u_1u_2 \ldots u_kX_qv_k \ldots v_2v_1.$$

Similar as in the induction basis, we see that $r \in [(q \cdot u_{k+1}) \circ v_{k+1}]$ if and only if there is a derivation $X_q \Rightarrow_G^* u_{k+1}X_rv_{k+1}$. Combining these derivations leads to

$$X_p \Rightarrow_G^* u_1u_2 \ldots u_kX_qv_k \ldots v_2v_1 \Rightarrow_G^* u_1u_2 \ldots u_ku_{k+1}X_rv_{k+1}v_k \ldots v_2v_1.$$

Therefore, we have $r \in [((\ldots((((p \cdot u_1) \circ v_1) \cdot u_2) \circ v_2) \ldots) \cdot u_{k+1}) \circ v_{k+1}]$ if and only if $X_p \Rightarrow^* u_1u_2 \ldots u_{k+1}X_rv_{k+1} \ldots v_2v_1$. This proves the stated claim, and concludes our proof of the theorem. □

It follows directly from Theorem 7.5 that a language accepted by a nondeterministic biautomaton is linear context-free. The other conversion direction, from a linear context-free grammar in normal form to a nondeterministic biautomaton, is presented in the next result. The conversion is such that the constructed biautomaton is guaranteed to satisfy the F-property.

Theorem 7.6. *Let G be a linear context-free grammar in normal form with n nonterminals. Then one can effectively construct an $(n+1)$-state nondeterministic biautomaton A, such that $L(A) = L(G)$. Further, the biautomaton A can be forced to satisfy the F-property.*

Proof. Let $G = (N, T, P, S)$ be a linear context-free grammar in normal form with $|N| = n$. We construct nondeterministic biautomaton $A = (Q, \Sigma, \cdot, \circ, I, F)$ over the alphabet $\Sigma = T$, with states $Q = \{\, q_X \mid X \in N \,\} \cup \{q_f\}$, the union being disjoint, initial states $I = \{q_S\}$, set of final states $F = \{\, q_S \mid S \to \lambda \in P \,\} \cup \{q_f\}$, and transition functions $\cdot$ and $\circ$ such that for all $X \in N$ and $a \in T$ we have

$$q_X \cdot a = \{\, q_Y \mid X \to aY \in P \,\} \cup \{\, q_f \mid X \to a \in P \,\},$$
$$q_X \circ a = \{\, q_Y \mid X \to Ya \in P \,\} \cup \{\, q_f \mid X \to a \in P \,\}.$$

The transitions for state q_f are $q_f \cdot a = q_f \circ a = \emptyset$, for all $a \in T$. Observe that by construction A satisfies the F-property.

With an inductive argumentation similar as in the proof of Theorem 7.5 one can show that a word $w \in \Sigma^*$ is accepted by the biautomaton A if and only if it can be generated by the grammar G. We use a more interesting way to prove $L(A) = L(G)$ instead. We transform the biautomaton A back to a linear context-free grammar G' and show $L(G') = L(G)$. Using the technique from Theorem 7.5, we construct from automaton A the grammar $G' = (N', T, P', S')$ with nonterminals

$$N' = \{\, Z_q \mid q \in Q \,\} \cup \{S'\} = \{\, Z_{q_X} \mid X \in N \cup \{f\} \,\} \cup \{S'\}$$

and productions $P' = \{\, S \to \lambda \mid I \cap F \neq \emptyset \,\} \cup P'_{\mathrm{fwd}} \cup P'_{\mathrm{bwd}}$, where

$$P'_{\mathrm{fwd}} = \{\, S' \to aZ_{q_X} \mid q_X \in (I \cdot a) \,\} \cup \{\, S' \to a \mid (I \cdot a) \cap F \neq \emptyset \,\}$$
$$\cup \{\, Z_{q_X} \to aZ_{q_Y} \mid q_Y \in (q_X \cdot a) \,\} \cup \{\, Z_{q_X} \to a \mid (q_X \cdot a) \cap F \neq \emptyset \,\}$$

and

$$P'_{\mathrm{bwd}} = \{\, S' \to Z_{q_X}a \mid q_X \in (I \circ a) \,\} \cup \{\, S' \to a \mid (I \circ a) \cap F \neq \emptyset \,\}$$
$$\cup \{\, Z_{q_X} \to Z_{q_Y}a \mid q_Y \in (q_X \circ a) \,\} \cup \{\, Z_{q_X} \to a \mid (p \circ a) \cap F \neq \emptyset \,\}.$$

A direct correspondence between the productions of G' and those of G can be seen as follows. First, we have $S' \to \lambda \in P'$ if and only if $I \cap F \neq \emptyset$, which by the definitions of I and F holds if and only if $S \to \lambda \in P$. Now consider the productions in P_{fwd} and P'_{fwd}. For all nonterminals $X, Y \in N$, and terminals $a \in T$ we have

$$
\begin{aligned}
S' \to aZ_{q_X} \in P'_{\mathrm{fwd}} &\iff q_X \in (\{q_S\} \cdot a) &\iff S \to aX \in P_{\mathrm{fwd}}, \\
Z_{q_X} \to aZ_{q_Y} \in P'_{\mathrm{fwd}} &\iff q_Y \in (\{q_X\} \cdot a) &\iff X \to aY \in P_{\mathrm{fwd}}, \\
S' \to a \in P'_{\mathrm{fwd}} &\iff q_f \in (\{q_S\} \cdot a) &\iff S \to a \in P_{\mathrm{fwd}}, \text{ and} \\
Z_{q_X} \to a \in P'_{\mathrm{fwd}} &\iff q_f \in (\{q_X\} \cdot a) &\iff X \to a \in P_{\mathrm{fwd}},
\end{aligned}
$$

where for the left equivalences of the last two lines we use the fact that

$$(I \cdot a) \cap F = (\{q_S\} \cdot a) \cap (\{q_f\} \cup \{\, q_S \mid S \to \lambda \in P \,\}) = (\{q_S\} \cdot a) \cap \{q_f\},$$

which is true because of the following observation. Since grammar G is in normal form, the nonterminal S cannot occur on a right-hand side of a production if $S \to \lambda$ belongs to P. Therefore, if $q_S \in F$, then it cannot be $q_S \in (\{q_X\} \cdot a)$ for any nonterminal $X \in N$. With a similar argumentation one can see that the following holds for the productions in P_{bwd} and P'_{bwd}:

$$
\begin{aligned}
S' \to aZ_{q_X} \in P'_{\mathrm{bwd}} &\iff S \to aX \in P_{\mathrm{bwd}}, \\
Z_{q_X} \to aZ_{q_Y} \in P'_{\mathrm{bwd}} &\iff X \to aY \in P_{\mathrm{bwd}}, \\
S' \to a \in P'_{\mathrm{bwd}} &\iff S \to a \in P_{\mathrm{bwd}}, \text{ and} \\
Z_{q_X} \to a \in P'_{\mathrm{bwd}} &\iff X \to a \in P_{\mathrm{bwd}}.
\end{aligned}
$$

Note that since $q_f \cdot a = q_f \circ a = \emptyset$, the set P' does not contain any production with Z_{q_f} as left-hand side. With this relation between the productions of the grammars G and G', one sees that $L(G) = L(G')$. From the proof of Theorem 7.5 the equality $L(G') = L(A)$ follows, so we conclude $L(A) = L(G)$. $\square$

The following example illustrates the above given construction of a biautomaton from a linear context-free grammar in normal form.

Example 7.7. Recall the grammar $G' = (N', T, P', S)$ from Example 7.4 with the terminal alphabet $T = \{a, b, c\}$, set of nonterminals

$$N' = \{S, X, Y, [X_{bb}], [X_b], [{}_{aa}Y], [{}_aY]\},$$

the axiom S, and the following set of productions:

$$P' = \{\, S \to \lambda,\, S \to a[X_{bb}],\, [X_{bb}] \to [X_b]b,\, [X_b] \to Xb,\, X \to a[X_{bb}],\, X \to c \,\}$$
$$\cup \{\, S \to [{}_{aa}Y]b,\, [{}_{aa}Y] \to a[{}_aY],\, [{}_aY] \to aY,\, Y \to [{}_{aa}Y]b,\, [{}_aY] \to a \,\}.$$

If we use the construction from the proof of Theorem 7.6 to convert this linear context-free grammar G' into a biautomaton, we obtain the nondeterministic biautomaton $A = (Q, \Sigma, \cdot, \circ, I, F)$ over the input alphabet $\Sigma = T$, with state set

$$Q = \{q_S, q_X, q_Y, q_{[X_{bb}]}, q_{[X_b]}, q_{[{}_{aa}Y]}, q_{[{}_aY]}\} \cup \{q_f\},$$

initial state set $I = \{q_S\}$, and set of final states $F = \{q_f, q_S\}$—note that $S \to \lambda$ belongs to P'. The transition functions of A can be read from Figure 7.3 on the next page. These functions directly result from the set of productions P' of the grammar. For example we have $q_{[{}_aY]} \cdot a = \{q_Y, q_f\}$, because the two productions $[{}_aY] \to aY$ and $[{}_aY] \to a$ belong to P'. Notice that automaton A has the F-property. $\diamond$

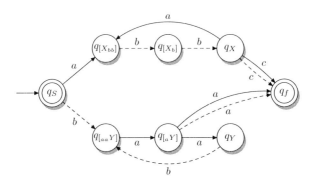

Figure 7.3: The nondeterministic biautomaton A with the F-property constructed from the linear context-free grammar G' in normal-form. The language accepted by A is $L(A) = \{\, a^i c b^{2i} \mid i \geq 1 \,\} \cup \{\, a^{2i} b^i \mid i \geq 0 \,\}$.

Theorems 7.5 and 7.6 imply that the class of languages accepted by nondeterministic biautomata coincides with the family of linear context-free languages. This even holds if we require the automata to satisfy the F-property. Remember that we have seen in Lemma 7.1 that any nondeterministic biautomaton can be transformed into a deterministic one, while preserving the structural properties. Therefore, we obtain a characterization of linear context-free languages in terms of different types of biautomata. In the following we write $\mathcal{L}(\mathcal{A})$, for an automaton type $\mathcal{A}$, to denote the class of languages accepted by automata of type $\mathcal{A}$. For example, the set of languages accepted by deterministic biautomata that satisfy the F-property is denoted by $\mathcal{L}(F\text{-DBiA})$. From above considerations we obtain the following result.

Corollary 7.8. *The class of linear context-free languages coincides with each of the four classes of languages accepted by deterministic or nondeterministic biautomata with or without the F-property, that is,*

$$\mathcal{L}(F\text{-DBiA}) = \mathcal{L}(F\text{-NBiA}) = \mathcal{L}(DBiA) = \mathcal{L}(NBiA) = LIN. \qquad \square$$

What can be said about languages accepted by deterministic or nondeterministic biautomata that satisfy the I-property? We will see in the proof of the upcoming result how to convert any biautomaton A into an equivalent biautomaton A', that satisfies the F-property, while preserving the I-property.

Lemma 7.9. *Let A be a deterministic or nondeterministic biautomaton. Then one can construct a nondeterministic biautomaton B satisfying the F-property, such that $L(B) = L(A)$. Moreover, if A has the I-property, then B has the I-property, too.*

Proof. Consider some biautomaton $A = (Q, \Sigma, \cdot, \circ, I, F)$, and construct the nondeterministic biautomaton $B = (Q_B, \Sigma, \cdot_B, \circ_B, I, F_B)$, where $Q_B = Q \cup \{q_f\}$

and $F_B = F \cup \{q_f\}$, for a new accepting state $q_f \notin Q$, and where the transition functions $\cdot_B$ and $\circ_B$ are defined such that for all states $q \in Q$ and input symbols $a \in \Sigma$ we have

$$q \cdot_B a = \begin{cases} (q \cdot a) \cup \{q_f\} & \text{if } (q \cdot a) \cap F \neq \emptyset \text{ or } (q \circ a) \cap F \neq \emptyset, \\ q \cdot a & \text{otherwise,} \end{cases}$$

$$q \circ_B a = \begin{cases} (q \circ a) \cup \{q_f\} & \text{if } (q \cdot a) \cap F \neq \emptyset \text{ or } (q \circ a) \cap F \neq \emptyset, \\ q \circ a & \text{otherwise.} \end{cases}$$

For the new state q_f we define $q_f \cdot a = q_f \circ a = \emptyset$, for all symbols $a \in \Sigma$. One can see that the biautomaton B has the F-property, because both sets $q \cdot_B a$ and $q \circ_B a$ contain the accepting state q_f if and only if at least one of the sets $q \cdot a$ and $q \circ b$ contains an accepting state of the biautomaton A. Further, since the newly added state q_f has no outgoing transitions, the language of the automaton has not changed, that is, $L(B) = L(A)$.

Now assume that the biautomaton A has the I-property, that is, $I \cdot a = I \circ a$, for all symbols $a \in \Sigma$. If $(I \cdot a) \cap F \neq \emptyset$, then we get

$$I \cdot_B a = (I \cdot a) \cup \{q_f\} = (I \circ a) \cup \{q_f\} = I \circ_B a$$

and otherwise we have $I \cdot_B a = I \cdot a = I \circ a = I \circ_B a$. Therefore, biautomaton B has the I-property, too. □

By Lemma 7.1 we know that any nondeterministic biautomaton can be determinized such that its structural properties are preserved. Together with the previous Lemma 7.9, we see that the families of languages accepted by deterministic or nondeterministic biautomata that have the I-property and do or do not have the F-property are all equal. The next result compares this language family with the classes of linear context-free language and regular languages.

Lemma 7.10. *The following relations between language classes hold.*

$$\mathcal{L}(F\text{-}I\text{-}DBiA) = \mathcal{L}(F\text{-}I\text{-}NBiA) = \mathcal{L}(I\text{-}DBiA) = \mathcal{L}(I\text{-}NBiA) \subsetneq LIN.$$

Moreover, the language classes $\mathcal{L}(F\text{-}I\text{-}DBiA)$ and REG are incomparable.

Proof. The fact that the four language classes of biautomata with the I-property are equal follows from Lemmas 7.1 and 7.9. Inclusion in LIN follows from the fact that $\mathcal{L}(I\text{-}NBiA) \subseteq \mathcal{L}(NBiA)$ and by Corollary 7.8. The strictness of the inclusion $\mathcal{L}(F\text{-}I\text{-}DBiA) \subsetneq LIN$ follows from the incomparability between $\mathcal{L}(I\text{-}NBiA)$ and REG, which will be shown in the following.

First consider the nondeterministic biautomaton $A = (Q, \Sigma, \cdot, \circ, I, F)$, which is depicted in Figure 7.4 on the following page: the automaton has the set of states $Q = \{q_0, q_1, q_2\}$, initial state set $I = \{q_0\}$, accepting states $F = \{q_1\}$, and the following transitions:

$$q_0 \cdot c = q_0 \circ c = \{q_1\}, \qquad q_1 \cdot a = \{q_2\}, \qquad q_2 \circ b = \{q_1\},$$

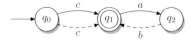

Figure 7.4: The nondeterministic biautomaton A with the I-property, that accepts the non-regular language $L(A) = \{\, ca^n b^n, a^n b^n c \mid n \geq 0 \,\}$.

and all undefined transitions map to the empty set. Notice that A has the I-property. Moreover, it is not hard to see that the language accepted by this biautomaton is $L(A) = \{\, ca^n b^n, a^n b^n c \mid n \geq 0 \,\}$, which is not a regular language. Therefore $\mathcal{L}(I\text{-}F\text{-DBiA}) = \mathcal{L}(I\text{-NBiA}) \not\subseteq \text{REG}$.

For the other non-inclusion, namely $\text{REG} \not\subseteq \mathcal{L}(I\text{-DBiA}) = \mathcal{L}(I\text{-}F\text{-DBiA})$, we argue as follows. If $A = (Q, \Sigma, \cdot, \circ, q_0, F)$ is a deterministic biautomaton with the I-property, then for all symbols $a, b \in \Sigma$ and words $v \in \Sigma^*$, the word avb is accepted by A if and only if there is a state $q \in Q$ with $q_0 \cdot a = q$ and $vb \in L_A(q)$, or there is a state $p \in Q$ with $q_0 \circ b = p$ and $av \in L_A(p)$. Because A has the I-property, in the first case we also have $q_0 \circ a = q$, so the automaton also accepts the word vba. Similarly, in the second case we have $q_0 \cdot b = p$, so the automaton accepts the word bav. This shows that every language $L \subseteq \Sigma^*$ in the class $\mathcal{L}(I\text{-}F\text{-DBiA})$ satisfies the following property:

for all $a, b \in \Sigma$ and $v \in \Sigma^*$, if $avb \in L$, then $\{vba, bav\} \cap L \neq \emptyset$.

Obviously, this property is not satisfied for the regular language $\{abc\}$, so we see that $\text{REG} \not\subseteq \mathcal{L}(I\text{-}F\text{-DBiA})$. $\qquad\square$

We have investigated the F-property and the I-property. The third of the structural properties for biautomata which were defined in Section 6.2, the $\diamond$-property, is a much stronger restriction than the other two properties. We will see in the next section that automata with the $\diamond$-property accept only regular languages.

7.2   Regular Languages

We have already seen in Section 6.2.1 that the family of languages accepted by deterministic or nondeterministic biautomata that satisfy both the $\diamond$-property and the F-property coincides with the class of regular languages. By Lemma 7.1 we know that the same is true if we consider nondeterministic biautomata instead of deterministic ones. But in the following we will see that the F-property is not essential here, since already the $\diamond$-property alone guarantees that the language accepted by such a biautomaton is regular.

Theorem 7.11. *Let A be a nondeterministic biautomaton with the $\diamond$-property. Then $L(A)$ is a regular language.*

Proof. Let $A = (Q, \Sigma, \cdot, \circ, I, F)$ be a biautomaton that satisfies the $\diamond$-property. Corollary 6.6 implies that A accepts a word $w \in \Sigma^*$ if and only if there are words $u, v \in \Sigma^*$, with $w = uv$ and $[(I \cdot u) \circ v] \cap F \neq \emptyset$. This means that there are states $q_0, q, q_f \in Q$, such that $q_0 \in I$, $q \in q_0 \cdot u$, $q_f \in q \circ v$, and $q_f \in F$. Thus, the language $L(A)$ can be described by

$$L(A) = \bigcup_{q \in Q} \{ u \mid q \in I \cdot u \} \cdot \{ v \mid (q \circ v) \cap F \neq \emptyset \}.$$

It remains to show that for every state $q \in Q$ the sets $L_1(q) = \{ u \mid q \in I \cdot u \}$ and $L_2(q) = \{ v \mid (q \circ v) \cap F \neq \emptyset \}$ are regular. The language $L_1(q)$ consists of all the words that, when read forward, lead from some initial state to state q, while language $L_2(q)$ consists of the words that, when read backwards, lead from state q to some accepting state. Thus, language $L_1(q)$ is accepted by the nondeterministic finite automaton $A_1 = (Q, \Sigma, \delta_1, I, \{q\})$ with $\delta_1(p, a) = p \cdot a$, for all $p \in Q$ and $a \in \Sigma$, which shows that $L_1(q)$ is a regular language. Further, the language $L_2(q)^R$ is accepted by the nondeterministic finite automaton $A_2 = (Q, \Sigma, \delta_2, \{q\}, F)$, with $\delta_2(p, a) = p \circ a$, for every $p \in Q$ and $a \in \Sigma$. Since regular languages are closed under reversal, the set $L_2(q)$ is a regular language, too. Since regular languages are closed under concatenation and union, the proof is complete. □

7.2.1 Subregular Language Classes

Biautomata were used in two papers of Klíma and Polák [74, 75] to give alternative characterizations of well-known subregular language families, such as prefix-suffix testable languages and piecewise testable languages. Further biautomata characterizations for subregular language classes, like for example starfree languages and definite languages, were given by Holzer and Jakobi [54]. In this subsection we will consider further subregular language classes, namely cyclic, commutative, and unary regular languages.

Let L be a language over the alphabet Σ. The *cyclic shift* of L is the language

$$\circlearrowright(L) = \{ vu \in \Sigma^* \mid uv \in L \}.$$

A language L is *cyclic* if it is closed under cyclic shift, that is, if $\circlearrowright(L) = L$. Recall the statement of Lemma 6.12: if $A = (Q, \Sigma, \cdot, \circ, I, F)$ is a nondeterministic biautomaton that has the $\diamond$- and the I-property, then $(I \cdot uv) \circ w = (I \cdot v) \circ wu$ holds for all words $u, v, w \in \Sigma^*$. This means that the left language of every state $q \in Q$ is cyclic: if $q \in (I \cdot u) \circ v$, for some $u, v \in \Sigma^*$, then also $q \in I \cdot vu$. In particular the language accepted by A is cyclic, so we obtain the following result.

Corollary 7.12. *Let A be a nondeterministic biautomaton with both the $\diamond$-property and the I-property, then $L(A)$ is a regular cyclic language.* □

In fact, for the canonical biautomaton—see Section 6.2.1—also the converse implication holds. Thus, we obtain the following characterization of regular cyclic languages.

Theorem 7.13. *A regular language is cyclic if and only if its canonical biautomaton has the I-property.*

Proof. If A is a biautomaton with the $\diamond$- and the I-property, then we know from Corollary 7.12 that $L(A)$ is a regular cyclic language. For the converse implication, let $L \subseteq \Sigma^*$ be a regular cyclic language, and let $A_L = (Q, \Sigma, \cdot, \circ, \{q_0\}, F)$ be its canonical biautomaton with initial state $q_0 = L$—remember that the states of A_L are both-sided derivatives of the form $u^{-1}Lv^{-1}$, for $u, v \in \Sigma^*$. Since L is cyclic, for every word $v \in \Sigma^*$ and symbol $a \in \Sigma$ we have $av \in L$ if and only if $va \in L$. Thus, for every symbol $a \in \Sigma$, we obtain

$$q_0 \cdot a = a^{-1}L = \{\, v \in \Sigma^* \mid av \in L \,\} = \{\, v \in \Sigma^* \mid va \in L \,\} = La^{-1} = q_0 \circ a$$

Therefore, biautomaton A has the I-property, which proves the stated claim. $\square$

If a biautomaton has the I-property, then it does not matter whether a transition on some symbol $a \in \Sigma$ taken from an initial state is a forward or a backward transition. Now consider the stronger restriction that in every state $q \in Q$ it does not matter whether one takes the forward transition or the backward transition on symbol a. In other words, we require $q \cdot a = q \circ a$, for all states $q \in Q$ and symbols $a \in \Sigma$. In this way we obtain a characterization of commutative regular languages by the structure of their canonical biautomaton. A language $L \subseteq \Sigma^*$ is *commutative* if for all words $u, v \in \Sigma^*$ and letters $a, b \in \Sigma$ we have $uabv \in L$ if and only if $ubav \in L$. One can see by induction that this condition is equivalent to the condition that for all words $u, v, x, y \in \Sigma^*$ we have $uxyv \in L$ if and only if $uyxv \in L$.

Theorem 7.14. *Let $L \subseteq \Sigma^*$ be a regular language and $A = (Q, \Sigma, \cdot, \circ, I, F)$ its canonical biautomaton. Then L is commutative if and only if $q \cdot a = q \circ a$, for every state $q \in Q$ and symbol $a \in \Sigma$.*

Proof. Let $L \subseteq \Sigma^*$ be a regular language, and let $A = (Q, \Sigma, \cdot, \circ, I, F)$ be the canonical biautomaton of L with $q \cdot a = q \circ a$, for all $q \in Q$ and $a \in \Sigma$. Then, by Corollary 7.12, all the right languages $L(_qA)$ of states $q \in Q$ are cyclic, because $_qA$, the automaton obtained by making q the initial state of A, has the I-property. Since the right languages of states of the canonical biautomaton are quotients $u^{-1}Lv^{-1}$, for $u, v \in \Sigma^*$, all these quotients are cyclic. This means that for all $u, v, x, y \in \Sigma^*$ we have $xy \in u^{-1}Lv^{-1}$ if and only if $yx \in u^{-1}Lv^{-1}$. It follows that L is commutative.

For the converse, assume L is commutative. Consider a letter $a \in \Sigma$, and a state $q \in Q$ that corresponds to a quotient $u^{-1}Lv^{-1}$, for some words $u, v \in \Sigma^*$.

We have

$$q \cdot a = [u^{-1}Lv^{-1}] \cdot a = (ua)^{-1}Lv^{-1} = \{\, x \in \Sigma^* \mid uaxv \in L \,\}$$

and

$$q \circ a = [u^{-1}Lv^{-1}] \circ a = u^{-1}L(av)^{-1} = \{\, x \in \Sigma^* \mid uxav \in L \,\},$$

and since L is commutative, it follows $q \cdot a = q \circ a$. $\square$

Note that the condition $q \cdot a = q \circ a$ together with the $\diamond$-property in particular implies that $(q \cdot a) \cdot b = (q \cdot b) \cdot a$ holds for the forward transition function of the canonical biautomaton. This nicely shows the connection to commutative finite automata [104], where $\delta(q, ab) = \delta(q, ba)$ holds for the transition function δ of the finite automaton.

We conclude this section with a remark on unary languages, which are languages defined over an alphabet consisting of a single symbol only. Seymour Ginsburg and H. Gordon Rice [39] have shown that every unary context-free language is regular. Since the language accepted by a biautomaton is linear context-free, any biautomaton with a unary input alphabet accepts a regular language, regardless which structural properties it has or has not. In fact, any unary biautomaton can be interpreted as a nondeterministic finite automaton in the following sense.

Fact 7.15. *Let* $A = (Q, \Sigma, \cdot, \circ, \{q_0\}, F)$ *be a biautomaton over the unary alphabet* $\Sigma = \{a\}$. *Then* $A' = (Q, \Sigma, \delta, q_0, F)$ *with* $\delta(q, a) = (q \cdot a) \cup (q \circ a)$, *for all* $q \in Q$ *and* $a \in \Sigma$, *is a nondeterministic finite automaton, with* $L(A') = L(A)$.

7.3 The Dual of a Biautomaton

For classical finite automata, an automaton for the reversal of the accepted language can be obtained by constructing the reversal, or dual automaton, that is, by reversing the transitions and interchanging initial and final states—see page 72 of Chapter 5. For biautomata, one obtains an automaton for the reversal of the language by simply interchanging the transition functions $\cdot$ and $\circ$. Nevertheless, it is interesting to see what happens if we apply a similar construction as for finite automata to biautomata. Furthermore, we will use some of the upcoming results in Section 9.2 to develop a minimization algorithm for $\diamond$-F-DBiAs, which is similar to Brzozowski's minimization algorithm for finite automata [12].

Let us define the *dual*[1] of the biautomaton $A = (Q, \Sigma, \cdot, \circ, I, F)$ as the biautomaton $A^R = (Q, \Sigma, \cdot^R, \circ^R, F, I)$ that is obtained from A by interchanging

---

[1]In the classical setting of finite automata, the described automaton is also called the *reversal* automaton, because the language it accepts is the reversal of the language accepted by the original finite automaton. Because, as we will see in the following, this relation between the accepted languages does *not* hold in the case of biautomata, we do not use the name reversal automaton here.

the initial and final states and by reversing all transitions, such that $p \in q \cdot^R a$ if and only if $q \in p \cdot a$, and $p \in q \circ^R a$ if and only if $q \in p \circ a$. Note that $(A^R)^R = A$.

The following lemma shows some useful reachability properties of the dual biautomaton.

Lemma 7.16. *Let $A = (Q, \Sigma, \cdot, \circ, I, F)$ be a nondeterministic biautomaton, and refer to the dual of A by $A^R = (Q, \Sigma, \cdot^R, \circ^R, F, I)$. Then for all states $p, q \in Q$ and words $u, v \in \Sigma^*$, we have $q \in (p \cdot u) \circ v$ if and only if $p \in (q \circ^R v^R) \cdot^R u^R$, and $q \in (p \circ v) \cdot u$ if and only if $p \in (q \cdot^R u^R) \circ^R v^R$.*

Proof. We prove the first part of the statement, namely $q \in (p \cdot u) \circ v$ if and only if $p \in (q \circ^R v^R) \cdot^R u^R$, by induction on the word $|uv|$. The second part of the statement then follows by symmetric argumentation, since $(A^R)^R = A$.

For the inductive basis $|uv| = 0$, that is, for $u = v = \lambda$, we have

$$(p \cdot u) \circ v = \{p\} = (p \circ^R v^R) \cdot^R u^R,$$

so the statement holds in this case.

Now let $|uv| \geq 1$, so we have $uv = u'av'$, with $u', v' \in \Sigma^*$ and $a \in \Sigma$, such that either $u = u'a$ and $v = v'$, or $u = u'$ and $v = av'$. First consider the case $u = u'a$. Then we have $q \in (p \cdot u) \circ v$ if and only if there are states p_1 and p_2, such that $p_1 \in p \cdot u'$, $p_2 \in p_1 \cdot a$, and $q \in p_2 \circ v$. Now we can apply the inductive assumption on the words u' and v, since both are shorter than $uv = u'av'$, and see that $p_1 \in p \cdot u'$ holds if and only if $p \in p_1 \cdot^R u'^R$, as well as $q \in p_2 \circ v$ if and only if $p_2 \in q \circ^R v^R$. Further, by definition of $\cdot^R$, we have $p_2 \in p_1 \cdot a$ if and only if $p_1 \in p_2 \cdot^R a$. Putting all this together, we have $q \in (p \cdot u) \circ v$ if and only if there are states p_1 and p_2, such that $p_2 \in q \circ^R v^R$, $p_1 \in p_2 \cdot^R a$, and $p \in p_1 \cdot^R u'^R$, that is, if and only if $p \in (q \circ^R v^R) \cdot^R u^R$.

For the other case, where $v = av'$, we use a similar argumentation: we have $q \in (p \cdot u) \circ v = ((p \cdot u) \circ v') \circ a$ if and only if there is a state $p_1 \in Q$, such that $p_1 \in (p \cdot u) \circ v'$, and $q \in p_1 \circ a$. By the inductive assumption, we have $p_1 \in (p \cdot u) \circ v'$ if and only if $p \in (p_1 \circ^R v'^R) \cdot^R u^R$, and by the definition of $\circ^R$, we have $q \in p_1 \circ a$ if and only if $p_1 \in q \circ^R a^R$. Thus, we have $q \in (p \cdot u) \circ v$ if and only if $p \in ((q \circ^R a^R) \circ^R v'^R) \cdot^R u^R = (q \circ^R v^R) \cdot^R u^R$. □

Note that if A has the $\diamond$-property, then the statement of Lemma 7.16 can be simplified to $q \in (p \cdot u) \circ v$ if and only if $p \in (q \cdot^R u^R) \circ^R v^R$. Further, for biautomata with the $\diamond$-property, the lemma also covers longer computations of the form

$$((\dots((((q \cdot u_1) \circ v_1) \cdot u_2) \circ v_2) \dots) \cdot u_k) \circ v_k$$

where the movement changes between the heads more than one time. In fact Lemma 7.16 can also be generalized to such longer computations in biautomata without the $\diamond$-property, as the following result shows. Note that the next lemma also covers the case where the computation in A starts with backward transitions, because the word u_1 in the statement of the lemma may be the empty word.

Lemma 7.17. *Let* $A = (Q, \Sigma, \cdot, \circ, I, F)$ *be a nondeterministic biautomaton, and refer to the dual of* A *by* $A^R = (Q, \Sigma, \cdot^R, \circ^R, F, I)$. *Then for all states* p *and* q *and words* $u_1, v_1, u_2, v_2, \ldots, u_k, v_k \in \Sigma^*$, *we have*

$$q \in ((\ldots ((((p \cdot u_1) \circ v_1) \cdot u_2) \circ v_2) \ldots) \cdot u_k) \circ v_k$$

if and only if

$$p \in ((((\ldots ((q \circ^R v_k^R) \cdot^R u_k^R) \ldots) \circ^R v_2^R) \cdot^R u_2^R) \circ^R v_1^R) \cdot^R u_1^R.$$

Proof. We prove the statement by induction on k. The induction basis for $k = 1$ is proven by Lemma 7.16. For the inductive step, assume that the statement is true for some integer k, and consider the following computation in the biautomaton A:

$$q \in ((((\ldots ((((p \cdot u_1) \circ v_1) \cdot u_2) \circ v_2) \ldots) \cdot u_k) \circ v_k) \cdot u_{k+1}) \circ v_{k+1},$$

for words $u_1, v_1, u_2, v_2, \ldots, u_{k+1}, v_{k+1} \in \Sigma^*$ and two states $p, q \in Q$. This relation between the states p and q holds if and only if there exists a state $r \in Q$ that satisfies

$$r \in ((((\ldots ((((p \cdot u_1) \circ v_1) \cdot u_2) \circ v_2) \ldots) \cdot u_k) \circ v_k) \quad \text{and} \quad q \in (r \cdot u_{k+1}) \circ v_{k+1}.$$

By applying the inductive assumption to the condition on the left and by using Lemma 7.16 on the condition on the right, we see that this is equivalent to the fact that there is a state $r \in Q$, which satisfies

$$p \in ((((\ldots ((r \circ^R v_k^R) \cdot^R u_k^R) \ldots) \circ^R v_2^R) \cdot^R u_2^R) \circ^R v_1^R) \cdot^R u_1^R$$

and $r \in (q \circ^R v_{k+1}^R) \cdot^R u_{k+1}^R$. This in turn holds if and only if

$$p \in ((((\ldots ((((q \circ^R v_{k+1}^R) \cdot^R u_{k+1}^R) \circ^R v_k^R) \cdot^R u_k^R) \ldots) \circ^R v_2^R) \cdot^R u_2^R) \circ^R v_1^R) \cdot^R u_1^R.$$

This concludes our proof. $\qquad\square$

The following result shows how the structural properties of the dual biautomaton are related to those of the original biautomaton: the $\diamond$-property is preserved, while the F- and the I-property are interchanged.

Lemma 7.18. *Let* A *be a nondeterministic biautomaton, then the following holds:*

1. *A has the $\diamond$-property if and only if A^R has the $\diamond$-property.*

2. *A has the F-property if and only if A^R has the I-property.*

3. *A has the I-property if and only if A^R has the F-property.*

Proof. Let $A = (Q, \Sigma, \cdot, \circ, I, F)$ be a biautomaton and $A^R = (Q, \Sigma, \cdot^R, \circ^R, F, I)$ be its dual. Assume A has the $\diamond$-property, so $(q \cdot a) \circ b = (q \circ b) \cdot a$, for every $q \in Q$ and $a, b \in \Sigma$. To see that also A^R has the $\diamond$-property, consider a state $q \in Q$, and note that by Lemma 7.16 we have $(q \cdot^R a) \circ^R b = \{\, p \in Q \mid q \in (p \circ b) \cdot a \,\}$. Since A has the $\diamond$-property, we know that

$$\{\, p \subset Q \mid q \subset (p \circ b) \cdot a \,\} = \{\, p \subset Q \mid q \subset (p \cdot a) \circ b \,\},$$

and by using Lemma 7.16 again, we obtain $\{\, p \in Q \mid q \in (p \cdot a) \circ b \,\} = (q \circ^R b) \cdot^R a$. Thus, we have shown $(q \cdot^R a) \circ^R b = (q \circ^R b) \cdot^R a$, for every $q \in Q$ and $a, b \in \Sigma$. So the $\diamond$-property of A implies the $\diamond$-property of A^R. The reverse implication immediately follows, since $(A^R)^R = A$, so the first statement is proven.

Now we show that the F-property of A implies the I-property of A^R. If A has the F-property, then $\{\, q \in Q \mid (q \cdot a) \cap F \neq \emptyset \,\} = \{\, q \in Q \mid (q \circ a) \cap F \neq \emptyset \,\}$, for every symbol $a \in \Sigma$. Notice that by Lemma 7.16 we have

$$F \cdot^R a = \bigcup_{f \in F} f \cdot^R a = \bigcup_{f \in F} \{\, q \in Q \mid f \in q \cdot a \,\} = \{\, q \in Q \mid (q \cdot a) \cap F \neq \emptyset \,\}, \text{ and}$$

$$F \circ^R a = \bigcup_{f \in F} f \circ^R a = \bigcup_{f \in F} \{\, q \in Q \mid f \in q \circ a \,\} = \{\, q \in Q \mid (q \circ a) \cap F \neq \emptyset \,\}.$$

Therefore, the dual biautomaton A^R has the I-property—note that F is the set of initial states of A^R.

For the reverse implication, assume that A^R has the I-property. Then for all states $q \in Q$ and input symbols $a \in \Sigma$ we have $q \in F \cdot^R a$ if and only if $q \in F \circ^R a$. Notice that by definition of $\cdot^R$ and $\circ^R$, we have $q \in F \cdot^R a$ if and only if $(q \cdot a) \cap F \neq \emptyset$, and similarly, $q \in F \circ^R a$ if and only if $(q \circ a) \cap F \neq \emptyset$. This shows that the I-property of the dual biautomaton A^R also implies the I-property for the biautomaton A. Thus, the second statement is proven. The final statement now follows from the fact that $(A^R)^R = A$. $\qquad\square$

If we have a biautomaton A with both the $\diamond$-property and the F-property, then Lemma 7.18 implies that the dual of the biautomaton A has the $\diamond$-property and the I-property. Then we know by Corollary 7.12 that the language accepted by the dual is a cyclic language. In fact, we can show the following result on the relation between the languages accepted by a biautomaton and its dual.

Corollary 7.19. *Let $A = (Q, \Sigma, \cdot, \circ, I, F)$ be a nondeterministic biautomaton that has both the $\diamond$- and the F-property and let $A^R = (Q, \Sigma, \cdot^R, \circ^R, F, I)$ be the dual of A. Then $L(A^R) = \circlearrowleft(L(A)^R)$, that is, automaton A^R accepts the cyclic shift of the reversal of $L(A)$.*

Proof. Let $w \in L(A)$. By Corollary 6.11, we have $(I \cdot w) \cap F \neq \emptyset$, so there is a state $p \in I$, and a state $q \in F$, such that $q \in p \cdot w$. By Lemma 7.16, this is equivalent to $p \in q \cdot^R w^R$, which means that w^R is accepted by A^R. Since by Corollary 7.12, the language $L(A^R)$ is cyclic, we know that $\circlearrowleft(w^R) \subseteq L(A^R)$, for every $w \in L(A)$, so $\circlearrowleft(L(A)^R) \subseteq L(A^R)$ holds.

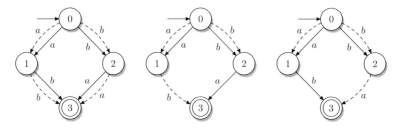

Figure 7.5: Three different nondeterministic biautomata A_1 (left), A_2 (middle), and A_3 (right), that have the $\diamond$-property and the I-property. All three automata accept the language $\{ab, ba\}$.

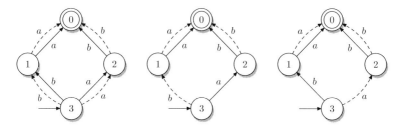

Figure 7.6: The dual automata A_1^R (left), A_2^R (middle), and A_3^R (right) of the corresponding biautomata from Figure 7.5, accepting pairwise different languages: $L(A_1^R) = \{ab, ba\}$, $L(A_2^R) = \{ab\}$, and $L(A_3)^R = \{ba\}$.

For the other inclusion let $w \in L(A^R)$. Recall that the automaton A^R has the $\diamond$-property, because A has this property. Therefore we know by Corollary 6.6 that there are words $u, v \in \Sigma^*$, with $uv = w$, such that $q_0 \in (q_f \cdot^R u) \circ^R v$, for some $q_f \in F$, and $q_0 \in I$. Lemma 6.12 implies that also $q_0 \in q_f \cdot^R vu$, which is equivalent to $q_f \in q_0 \cdot^R (vu)^R$, by Lemma 7.16. This means that the word $(vu)^R$ is accepted by the biautomaton A, so $vu \in L(A)^R$. Since $w = uv \in \circlearrowleft(vu)$, we have $w \in \circlearrowleft(L(A)^R)$, which proves $\circlearrowleft(L(A)^R) \supseteq L(A^R)$. $\square$

What can be said about $L(A^R)$, if A is a biautomaton with both the $\diamond$- and the I-property? Unfortunately, the language cannot be identified by some operation on the language $L(A)$, because a regular cyclic language can be accepted by structurally different biautomata that have the $\diamond$- and the I-property. From these structural differences, also different dual automata, that accept different languages, can result. This is can be observed in the following example.

Example 7.20. Consider the cyclic language $L = \{ab, ba\}$, which is accepted by all the three biautomata A_1, A_2, and A_3 that are depicted in Figure 7.5. Notice that these three automata have both the $\diamond$- and the I-property. Their corresponding dual automata A_1^R, A_2^R, and A_3^R are depicted in Figure 7.6. Note that these three automata have both the $\diamond$- and the F-property. Further, notice that the languages $L(A_1^R) = \{ab, ba\}$, $L(A_2^R) = \{ab\}$, and $L(A_3)^R = \{ba\}$ accepted by the dual automata are pairwise distinct. $\diamond$

Chapter 8

Descriptional Complexity of Conversions for Biautomata

OR comparing the biautomaton model with other descriptional systems for regular languages, this chapter is devoted to the study descriptional complexity aspects of biautomata. A basic question in descriptional complexity theory is, how succinctly a language can be described using a certain descriptional model, compared to other models. If a language can be described by two different models, one is interested how costly it is, in terms of the size of the description, to replace the representation of the language in one model by a representation of that language in the other model. Perhaps the most prominent example for such trade-offs is the conversion of a nondeterministic finite automaton into a deterministic one: Rabin and Scott [105] showed that any NFA with n states can be transformed into a DFA accepting the same language, such that this DFA has at most 2^n states. It is well known that in general this bound cannot be reduced, since there are n-state NFAs, for which every equivalent minimal DFA needs 2^n states—see for example the papers of Meyer and Fischer [94], and Moore [97].

Descriptional complexity aspects of biautomata were first studied by Galina Jiráskova and Ondřej Klíma [68]. There, tight bounds for conversions from deterministic and nondeterministic finite automata, as well as from syntactic monoids to biautomata with the ◇- and the F-property were shown. We investigate similar conversions for nondeterministic biautomata with both these properties. The descriptional complexity of conversions from deterministic and nondeterministic finite automata to nondeterministic biautomata is analyzed in Section 8.1, while Section 8.3 deals with the conversion from syntactic monoids. In between, in Section 8.2 we also consider regular expressions. Moreover, Section 8.4 studies descriptional complexity aspects of biautomata that do not have both the ◇-property and the F-property.

We start our investigations with the classical problem of determinization, that is, the conversion of nondeterministic to deterministic biautomata. The following upper bound follows directly from the power-set construction from Lemma 7.1.

Corollary 8.1. *For any given nondeterministic biautomaton A with n states, one can construct an equivalent deterministic biautomaton B having at most 2^n states, such that for all $X \in \{\diamond, F, I\}$, if A has the X-property, then B has the X-property, too.* □

An exponential lower bound for the cost of determinizing biautomata with $\diamond$- and F-property is given in the following result.

Lemma 8.2. *For all integers $n \geq 1$ there is a binary regular language L_n accepted by a nondeterministic biautomaton with $\diamond$-property and F-property that has $3n+2$ states, and for which every equivalent deterministic biautomaton with $\diamond$- and F-property needs at least $2^{2n}+1$ states.*

Proof. Let $n = k+1 \geq 1$ and consider the language $L_n = \Sigma^* a \Sigma^{n-1} a \Sigma^*$ over the alphabet $\Sigma = \{a, b\}$. This language is accepted by the finite automaton $A = (Q, \Sigma, \delta, q_0, F)$, with state set $Q = \{\, q \mid 0 \leq q \leq k \,\} \cup \{q_0, q_f\}$, set of final states $F = \{q_f\}$, and the transition function δ of which is depicted in Figure 8.1. We first show that any deterministic biautomaton for the language L_n that has the $\diamond$- and the F-property needs $2^{2n}+1$ states. Therefor we use a combined product and power-set construction as presented by Jirásková and Klíma [68] to convert the finite automaton A into an equivalent deterministic biautomaton that has the $\diamond$-property and the F-property. This construction yields the biautomaton $B = (Q_B, \Sigma, \cdot_B, \circ_B, \{q_0^B\}, F_B)$ with state set $Q_B = \{\, (S, T) \mid S, T \subseteq Q \,\}$, initial state $q_0^B = (\{q_0\}, \{q_f\})$, final states $F_B = \{\, (S, T) \in Q_B \mid S \cap T \neq \emptyset \,\}$, and where the transition functions $\cdot_B$ and $\circ_B$ are defined as follows, for all states $(S, T) \in Q_B$ and inputs $c \in \Sigma$:

$$(S, T) \cdot_B c = (\delta(S, c), T), \quad \text{and} \quad (S, T) \circ_B c = (S, \delta^R(T, c)),$$

with $\delta^R(T, c) = \{\, p \in Q \mid \delta(p, c) \cap T \neq \emptyset \,\}$. Jirásková and Klíma [68] showed that this biautomaton has the $\diamond$- and F-property. We now show that the equivalent minimal deterministic biautomaton C' with both these properties has at least $2^{2n}+1$ states. Similar as for DFAs, this can be done by proving that there are this number of reachable and pairwise distinguishable states [68]. We consider states of the form $(\{q_0\} \cup S, \{q_f\} \cup T)$, with $S, T \subseteq \{0, 1, \ldots, k\}$, and

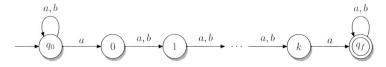

Figure 8.1: The nondeterministic finite automaton A over alphabet $\Sigma = \{a, b\}$ accepting the language $L_n = \Sigma^* a \Sigma^{n-1} a \Sigma^*$; recall that $n = k+1$.

state $(\{q_0, q_f\}, \{q_f\})$. The latter state can be reached from the initial state by reading $a^{k+2}b^{k+1}$ with forward transitions. Further, one can see by induction on the cardinality of $S \subseteq \{0, 1, \ldots, k\}$ that all states of the form $(\{q_0\} \cup S, \{q_f\})$ are reachable: for $|S| = \emptyset$, this is the initial state, and for $S = \{i_0, i_1, \ldots, i_\ell\}$, with $0 \leq i_0 < i_1 < \cdots < i_\ell \leq k$, we have

$$(\{q_0\} \cup S, \{q_f\}) = (\{i_1 - i_0 - 1, \ldots, i_\ell - i_0 - 1\}, \{q_f\}) \cdot_B ab^{i_0}.$$

Since the transition functions $\cdot_B$ and $\circ_B$ operate independently on the first and, respectively, second component of the states, also the state $(\{q_0\} \cup S, \{q_f\} \cup T)$ is reachable from the state $(\{q_0\} \cup S, \{q_f\})$, which can be seen by a similar inductive argument—notice that the NFA A is isomorphic to its reversal. To prove all considered states to be distinguishable, first note that the state $(\{q_0, q_f\}, \{q_f\})$ can be distinguished from every state $(\{q_0\} \cup S, \{q_f\} \cup T)$, with $S, T \subseteq \{0, 1, \ldots, k\}$, by reading b^{k+1} forward and reading b^{k+1} backward:

$$(\{q_0, q_f\}, \{q_f\}) \cdot_B b^{k+1} \circ_B b^{k+1} = (\{q_0, q_f\}, \{q_f\}) \in F_B,$$

$$(\{q_0\} \cup S, \{q_f\} \cup T) \cdot_B b^{k+1} \circ_B b^{k+1} = (\{q_0\}, \{q_f\}) \notin F_B.$$

Now consider two distinct states $(\{q_0\} \cup S, \{q_f\} \cup T)$ and $(\{q_0\} \cup S', \{q_f\} \cup T')$, with $S, S', T, T' \subseteq \{0, 1, \ldots, k\}$. If $S \neq S'$, then we may assume that there is an element $q \in S \setminus S'$. Then these states can be distinguished as follows:

$$(\{q_0\} \cup S, \{q_f\} \cup T) \cdot_B b^{k-q}ab^{k+1} \circ_B b^{k+1} = (\{q_0, q_f\}, \{q_f\}) \in F_B,$$

$$(\{q_0\} \cup S', \{q_f\} \cup T') \cdot_B b^{k-q}ab^{k+1} \circ_B b^{k+1} = (\{q_0\}, \{q_f\}) \notin F_B.$$

If otherwise $S = S'$, then it must be $T \neq T'$ and we may assume that there is an element $q \in T \setminus T'$. In this case the states can be distinguished as follows:

$$(\{q_0\} \cup S, \{q_f\} \cup T) \circ_B b^{k+1}ab^{k-q} \cdot_B b^{k+1} = (\{q_0\}, \{q_0, q_f\}) \in F_B,$$

$$(\{q_0\} \cup S', \{q_f\} \cup T') \circ_B b^{k+1}ab^{k-q} \cdot_B b^{k+1} = (\{q_0\}, \{q_f\}) \notin F_B.$$

Thus, any deterministic biautomaton with $\diamond$- and F-property that accepts L_n has at least $2^{2n} + 1$ states.

It remains to present a nondeterministic biautomaton with $\diamond$- and F-property for the language L_n that has $3n + 2$ states. Let $C = (Q_C, \Sigma, \cdot_C, \circ_C, I_C, F_C)$ be the nondeterministic biautomaton with state set

$$Q_C = \{ q, q', q'' \mid 0 \leq q \leq k \} \cup \{q_0, q_f\},$$

initial states $I_C = \{q_0\}$, final states $F_C = \{q_f, k''\}$, and the transition functions $\cdot_C$ and $\circ_C$, which are depicted in Figure 8.2 on the next page. One can see that C has $3n + 2$ states, and that it has the F-property—note that the

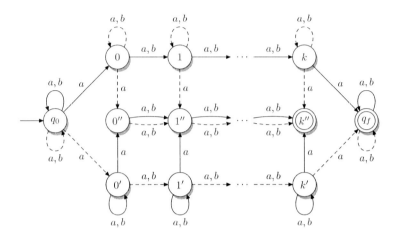

Figure 8.2: The nondeterministic biautomaton C over alphabet $\Sigma = \{a, b\}$ accepting the language $L_n = \Sigma^* a \Sigma^{n-1} a \Sigma^*$; recall that $n = k + 1$.

accepting states are q_f and k''. Next let us check the $\diamond$-property. For the initial state q_0 we have

$$(q_0 \cdot_C a) \circ_C a = \{q_0, 0\} \circ_C a = \{q_0, 0, 0', 0''\} = \{q_0, 0'\} \cdot_C a = (q_0 \circ_C a) \cdot_C a,$$
$$(q_0 \cdot_C a) \circ_C b = \{q_0, 0\} \circ_C b = \{q_0, 0\} = \{q_0\} \cdot_C a = (q_0 \circ_C b) \cdot_C a,$$
$$(q_0 \cdot_C b) \circ_C a = \{q_0\} \circ_C a = \{q_0, 0'\} = \{q_0, 0'\} \cdot_C b = (q_0 \circ_C a) \cdot_C b,$$
$$(q_0 \cdot_C b) \circ_C b = \{q_0\} \circ_C b = \{q_0\} = \{q_0\} \cdot_C b = (q_0 \circ_C b) \cdot_C b.$$

For states q, with $0 \le q \le k - 1$, we have

$$(q \cdot_C a) \circ_C a = \{(q+1)\} \circ_C a = \{(q+1), (q+1)''\} = \{q, q''\} \cdot_C a = (q \circ_C a) \cdot_C a,$$
$$(q \cdot_C a) \circ_C b = \{(q+1)\} \circ_C b = \{(q+1)\} = \{q\} \cdot_C a = (q \circ_C b) \cdot_C a,$$
$$(q \cdot_C b) \circ_C a = \{(q+1)\} \circ_C a = \{(q+1), (q+1)''\} = \{q, q''\} \cdot_C b = (q \circ_C a) \cdot_C b,$$
$$(q \cdot_C b) \circ_C b = \{(q+1)\} \circ_C b = \{(q+1)\} = \{q\} \cdot_C b = (q \circ_C b) \cdot_C b,$$

and for state k we have

$$(k \cdot_C a) \circ_C a = \{q_f\} \circ_C a = \{q_f\} = \{k, k''\} \cdot_C a = (k \circ_C a) \cdot_C a,$$
$$(k \cdot_C a) \circ_C b = \{q_f\} \circ_C b = \{q_f\} = \{k\} \cdot_C a = (k \circ_C b) \cdot_C a,$$
$$(k \cdot_C b) \circ_C a = \emptyset \circ_C a = \emptyset = \{k, k''\} \cdot_C b = (k \circ_C a) \cdot_C b,$$
$$(k \cdot_C b) \circ_C b = \emptyset \circ_C b = \emptyset = \{k\} \cdot_C b = (k \circ_C b) \cdot_C b.$$

The $\diamond$-property for states q', with $0 \le q \le k$, can be seen analogously to states q. For all states q'', with $0 \le q \le k - 2$ we have $(q'' \cdot_C c) \circ_C d = \{(q+2)''\}$, for $c, d \in \Sigma$. And for states $(k-1)''$ and k'' we have $((k-1)'' \cdot_C c) \circ_C d = \emptyset$ and

$(k'' \cdot_C c) \circ_C d = \emptyset$, for all $c, d \in \Sigma$. Finally, one readily sees that the $\diamond$-property also holds for state q_f.

Now that we know that biautomaton C has both the $\diamond$-property and the F-property, due to Corollary 6.11 it is easy to see that it accepts language L_n: a word $w \in \Sigma^*$ is accepted by C if and only if $(q_0 \cdot_C w) \cap F \neq \emptyset$. Hence, we only need to consider forward transitions of C and these correspond to the nondeterministic finite automaton A from Figure 8.1 on page 126. □

By setting $m = 3n + 2$, Lemma 8.2 gives a lower bound of $2^{(m-2)/3} + 1$ states for converting an m-state $\diamond$-F-NBiA to an equivalent $\diamond$-F-DBiA. It is left as an open question whether one can find some n-state nondeterministic biautomaton, for which any equivalent deterministic biautomaton needs 2^n states. In the following sections we consider conversions from different models for describing regular languages to biautomata with $\diamond$- and F-property. Descriptional complexity aspects of biautomata that do not have both these properties will be studied in Section 8.4.

8.1   From Finite Automata to Biautomata

The trade-off between finite automata and deterministic biautomata that satisfy the $\diamond$- and F-property was studied by Jirásková and Klíma [68]. They showed that every nondeterministic finite automaton with n states can be transformed into an equivalent deterministic biautomaton with $2^{2n} - 2 \cdot (2^n - 1)$ states, and this bound is tight. The corresponding tight bound for converting an n-state deterministic finite automaton into an equivalent deterministic biautomaton is $n \cdot 2^n - 2 \cdot (n - 1)$ states. Here we show that we can transform any finite automaton into an equivalent nondeterministic biautomaton with $\diamond$- and F-property of quadratic size. The following theorem provides the upper bound, which we later prove to be tight.

Theorem 8.3. *For any given (deterministic or nondeterministic) finite automaton with n states, one can construct an equivalent nondeterministic biautomaton with n^2 states that has the $\diamond$-property and the F-property.*

Proof. Given a finite automaton $A = (Q, \Sigma, \delta, I, F)$, let $B = (Q', \Sigma, \cdot, \circ, I', F')$ be the biautomaton with state set $Q' = Q \times Q$, set of initial states $I' = I \times F$, accepting states $F' = \{ (q, q) \mid q \in Q \}$, and where the transition functions $\circ$ and $\cdot$ are defined such that for all $(p, q) \in Q'$ and $a \in \Sigma$ we have

$$(p, q) \cdot a = \{ (p', q) \mid p' \in \delta(p, a) \}, \qquad (p, q) \circ a = \{ (p, q') \mid q \in \delta(q', a) \}.$$

Obviously, if A has n states, then B has n^2 states. Since the forward transition function $\cdot$ only operates on the first component of the states, and the backward transition function $\circ$ only operates on the second component, it follows that B has the $\diamond$-property. To see that it also has the F-property, notice that the set $(p, q) \circ a$ contains an accepting state from F', namely the element (p, p), if and only if we have $q \in \delta(p, a)$. But this in turn holds if and only if the set $(p, q) \cdot a$ contains an accepting state from F', namely the element (q, q). Thus, B is a biautomaton with the $\diamond$-property and the F-property. Then it is easy to see that $L(B) = L(A)$, since by Corollary 6.11 a word $w \in \Sigma^*$ is accepted by B if and only if there is an accepting computation that only uses forward transitions—and such computations correspond to computations in the original automaton A. $\qquad\square$

Our next goal is to prove a lower bound for this conversion. Therefor we use a straightforward adaptation of the extended fooling set technique for classical finite automata—see the paper of Birget [8]—to biautomata with the $\diamond$-property and the F-property.

Lemma 8.4. *A set $S = \{ (x_i, y_i, z_i) \mid x_i, y_i, z_i \in \Sigma^*, 1 \leq i \leq n \}$ is a bi-fooling set for a language $L \subseteq \Sigma^*$ if the following two properties hold:*

1. *for $1 \leq i \leq n$ we have $x_i y_i z_i \in L$, and*

2. *for $1 \leq i, j \leq n$, with $i \neq j$, we have $x_i y_j z_i \notin L$ or $x_j y_i z_j \notin L$.*

If S is a bi-fooling set for the language L, then any nondeterministic biautomaton with both the $\diamond$-property and the F-property that accepts the language L has at least $|S|$ states.

Proof. Let $S = \{ (x_i, y_i, z_i) \mid x_i, y_i, z_i \in \Sigma^*, 1 \leq i \leq n \}$ be a bi-fooling set for the language $L \subseteq \Sigma^*$ and let $A = (Q, \Sigma, \cdot, \circ, I, F)$ be a nondeterministic biautomaton that has the $\diamond$-property and the F-property, with $L(A) = L$. Then for all integers i, with $1 \leq i \leq n$, there exists $q_i \in Q$ such that $q_i \in (I \cdot x_i) \circ z_i$ and $(q_i \cdot y_i) \cap F \neq \emptyset$. If we assume for the sake of contradiction that $|Q| < |S|$, then there must exist two distinct integers i and j, with $1 \leq i, j \leq n$, such that $q_i = q_j$. But then it follows that $q_i \in (I \cdot x_i) \circ z_i$ and $q_i \in (I \cdot x_j) \circ z_j$, and moreover $(q_i \cdot y_i) \cap F \neq \emptyset$ and also $(q_i \cdot y_j) \cap F \neq \emptyset$. Therefore, the biautomaton A accepts both words $x_i y_j z_i$ and $x_j y_i z_j$, which is a contradiction to L being a bi-fooling set. So A must have at least $|S|$ states. $\qquad\square$

We now use the technique of Lemma 8.4 to prove a lower bound for the conversion from Theorem 8.3.

Theorem 8.5. *For all integers $n \geq 1$ there is a binary regular language L_n accepted by an n-state deterministic finite automaton, such that any nondeterministic biautomaton that has the $\diamond$-property and the F-property needs n^2 states to accept the language L_n.*

Proof. The statement for $n = 1$ can be seen with the witness language $\{a, b\}^*$. Now let $n \geq 2$ and consider the language L_n that is accepted by the deterministic finite automaton $A_n = (Q, \Sigma, \delta, q_0, F)$ over input alphabet $\Sigma = \{a, b\}$, with state set $Q = \{0, 1, \ldots, n-1\}$, initial state $q_0 = 0$, final states $F = \{0\}$, and the transition function δ of which is defined as follows—cf. Figure 8.3:

$$\delta(i,a) = \begin{cases} i+1 & \text{for } 0 \le i \le n-2, \\ 0 & \text{for } i = n-1, \end{cases} \qquad \delta(i,b) = \begin{cases} i+1 & \text{for } 0 \le i \le n-2, \\ n-1 & \text{for } i = n-1. \end{cases}$$

We now define a bi-fooling set of size n^2 for L_n, with $n \ge 3$—the case $n = 2$ will be discussed later. For $0 \le i, j \le n-1$ define the following words:

$$x_{i,j} = a^i, \qquad y_{i,j} = a^{n-i}b^{n-2}a^{j+2}, \qquad z_{i,j} = a^{n-j}.$$

Then the set $S = \{ (x_{i,j}, y_{i,j}, z_{i,j}) \mid 0 \le i,j \le n-1 \}$ is a bi-fooling set for L, which will be shown in the following. First note that for $0 \le i, j \le n-1$ the word

$$x_{i,j} \cdot y_{i,j} \cdot z_{i,j} = a^i \cdot a^{n-i}b^{n-2}a^{j+2} \cdot a^{n-j} = a^n b^{n-2} a^{n+2}$$

is accepted by A_n, thus, it belongs to the language L_n. Now consider pairs of integers (i,j), and (i',j'), with $0 \le i, i', j, j' \le n-1$, and $(i,j) \neq (i',j')$. We have to show that at least one of the words $x_{i,j}y_{i',j'}z_{i,j}$ and $x_{i',j'}y_{i,j}z_{i',j'}$ does not belong to the language L_n. We consider three cases.

First consider the case $i = i'$, where it must be $j \neq j'$. Then the word

$$x_{i,j} \cdot y_{i',j'} \cdot z_{i,j} = a^i \cdot a^{n-i'}b^{n-2}a^{j'+2} \cdot a^{n-j} = a^n b^{n-2} a^{n+2+j'-j}$$

is not accepted by the finite automaton A_n, which can be seen as follows. From the initial state 0, by reading $a^n b^{n-2}$, the automaton reaches state $n-2$. From there, a string a^m leads to the accepting state 0 if and only if $m \bmod n = 2$. But since $0 \le j, j' \le n-1$ and $j \neq j'$, we have $(n+2+j'-j) \bmod n \neq 2$—thus, the word is not accepted by A_n.

Second consider the case $i \neq i'$, and the sub-case where $(j'-j) \bmod n \neq n-1$ holds. We will see that the word

$$x_{i,j} \cdot y_{i',j'} \cdot z_{i,j} = a^i \cdot a^{n-i'}b^{n-2}a^{j'+2} \cdot a^{n-j} = a^{n-i'+i}b^{n-2}a^{n+2+j'-j}$$

is not accepted by A_n. After reading $a^{n-i'+i}$, the automaton A_n is in some state q, with $1 \le q \le n-1$, and then after reading b^{n-2} from state q, the automaton A_n reaches state $n-1$. From state $n-1$, a string a^m leads to the only accepting state 0 if and only if $m \bmod n = 1$. But from our assumption that $(j'-j) \bmod n \neq n-1$ holds, we conclude that $(n+2+j'-j) \bmod n \neq 1$, so the word cannot be accepted by A_n.

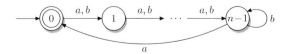

Figure 8.3: The deterministic finite automaton A_n for the lower bound of the conversion of finite automata to nondeterministic biautomata with $\diamond$-property and F-property.

Finally, consider the case $i \neq i'$, and $(j' - j) \bmod n = n - 1$. Here the word

$$x_{i',j'} \cdot y_{i,j} \cdot z_{i',j'} = a^{i'} \cdot a^{n-i} b^{n-2} a^{j+2} \cdot a^{n-j'} = a^{n-i+i'} b^{n-2} a^{n+2+j-j'}$$

is not accepted by A_n, which can be seen as follows. Again, after reading the first block of a symbols and the block of b symbols, the automaton A_n is in state $n - 1$. From there, the word $a^{n+2+j-j'}$ cannot lead to the accepting state 0, because

$$(n + 2 + j - j') \bmod n = (2 - ((j' - j) \bmod n)) \bmod n = (2 - n + 1) \bmod n \neq 1,$$

for $n \geq 3$. This concludes the proof that S is a bi-fooling set for L_n. Because $|S| = n^2$, the statement of the theorem for $n \geq 3$ follows from Lemma 8.4.

It remains to consider the case $n = 2$, where we use the following elements

$$(x_1, y_1, z_1) = (\lambda, \lambda, \lambda), \qquad (x_2, y_2, z_2) = (a, b, a),$$
$$(x_3, y_3, z_3) = (\lambda, a, a), \qquad (x_4, y_4, z_4) = (a, ba, \lambda),$$

to construct a bi-fooling set $S_2 = \{ (x_i, y_i, z_i) \mid 1 \leq i \leq 4 \}$ for $L_2 = L(A_2)$. To see that S_2 is a bi-fooling set for L_2, note that for $1 \leq i \leq 4$ the word $x_i y_i z_i$ belongs to L_2. Further, none of the words

$$x_1 y_2 z_1 = b, \qquad x_3 y_1 z_3 = a, \qquad x_4 y_1 z_4 = a,$$
$$x_2 y_3 z_2 = a^3, \qquad x_4 y_2 z_4 = ab, \qquad x_3 y_4 z_3 = ba^2$$

belongs to the language L_2. This concludes our proof. $\qquad\qquad\square$

8.2 From Regular Expressions to Biautomata

Prominent constructions for converting a regular expression into an equivalent nondeterministic finite automaton were given by Viktor M. Glushkov [40], and by Robert McNaughton and Hisao Yamada [92]. We show that a similar construction can be used to convert any regular expression into a nondeterministic biautomaton, that has both the $\diamond$- and the F-property. Given a regular expression r over an alphabet Σ, we denote by r' the regular expression, where all occurrences of alphabet symbols in r are numbered from left to right. For example, for $r = (a + b) \cdot a^*$, we have $r' = (a_1 + b_2) \cdot a_3^*$. The new alphabet over which r' is defined, is denoted by Γ. Further, let $\varphi \colon \Gamma \to \Sigma$ be the function that maps a numbered symbol $a_i \in \Gamma$ to its original symbol $a \in \Sigma$. Now we define the sets of positions of symbols that can appear at the beginning and at the end of words from $L(r')$,

$$\text{FIRST}_{r'} = \{ i \mid \exists v \in \Gamma^* : a_i v \in L(r') \},$$
$$\text{LAST}_{r'} = \{ i \mid \exists v \in \Gamma^* : v a_i \in L(r') \},$$

and for each position i, we define the set of positions of symbols that may follow symbol a_i:

$$\text{FOLLOW}_{r'}(i) = \{\, j \mid \exists u, v \in \Gamma^* : u a_i a_j v \in L(r') \,\}.$$

In the following we omit the subscript r' in the previous definitions, because the regular expressions r and r' are clear from the context. Gerard Berry and Ravi Sethi [7] showed how one can effectively construct the sets FIRST, LAST, and FOLLOW.

Let us define a biautomaton $A'_r = (Q_r, \Gamma, \cdot'_r, \circ'_r, I_r, F_r)$ accepting $L(r')$, with state set $Q_r = \{\, (i,j), (\bot, j), (i, \bot), (\bot, \bot) \mid a_i, a_j \in \Gamma \,\}$, final states

$$F_r = \{\, (\bot, j) \mid j \in \text{FIRST} \,\} \cup \{\, (i, \bot) \mid i \in \text{LAST} \,\}$$
$$\cup \{\, (i,j) \mid j \in \text{FOLLOW}(i) \,\} \cup \{\, (\bot, \bot) \mid \lambda \in L(r') \,\},$$

and initial states $I_r = \{\, (\bot, \bot) \,\}$. The transition functions $\cdot'_r$ and $\circ'_r$ are defined such that for all states $(x,y) \in Q_r$ and symbols $a_j \in \Gamma$, we have

$$(x,y) \cdot'_r a_j \ni (j,y) \text{ if } \begin{cases} \text{either } x = i \text{ for some } a_i \in \Gamma \text{ and } j \in \text{FOLLOW}(i), \\ \text{or } x = \bot \text{ and } j \in \text{FIRST}, \end{cases}$$

$$(x,y) \circ'_r a_j \ni (x,j) \text{ if } \begin{cases} \text{either } y = i \text{ for some } a_i \in \Gamma \text{ and } i \in \text{FOLLOW}(j), \\ \text{or } y = \bot \text{ and } j \in \text{LAST}. \end{cases}$$

To obtain a biautomaton for the language $L(r)$ we simply apply the mapping φ to the input symbols of A'_r: let $A_r = (Q_r, \Sigma, \cdot_r, \circ_r, I_r, F_r)$ be the biautomaton, where for all symbols $a_i \in \Gamma$ and states $p, q \in Q_r$ we have

$$p \cdot_r a = \bigcup_{\varphi(a_i)=a} p \cdot'_r a_i, \qquad p \circ_r a = \bigcup_{\varphi(a_i)=a} p \circ'_r a_i.$$

The following theorem shows the correctness of this construction and gives an upper bound on the size increase. Here the size of a regular expression r over an alphabet Σ is measured by the *alphabetic width* of r, which is the number of occurrences of symbols from Σ in the expression r. For example, the regular expression $(a + b) \cdot a^*$ has alphabetic width 3.

Theorem 8.6. *Let r be a regular expression of alphabetic width n. Then A_r is a nondeterministic biautomaton with $L(A_r) = L(r)$ that has $(n+1)^2$ states. Further, the biautomaton A_r has the $\diamond$- and the F-property.*

Proof. We first show that A'_r has both stated properties. Since the transition functions $\cdot'_r$ and $\circ'_r$ independently operate on the first and, respectively, second component of the states of A'_r, this biautomaton has the $\diamond$-property. To show that A'_r has the F-property, let $(x,y) \in Q_r$ and $a_i \in \Gamma$. We consider four cases, depending on the type of state (x,y), namely $(x,y) = (\bot, \bot)$, $(x,y) = (\bot, j)$, $(x,y) = (i, \bot)$, and $(x,y) = (i,j)$, for $a_i, a_j \in \Gamma$:

- If $(x, y) = (\bot, \bot)$, then $((x, y) \cdot'_r a_k) \cap F_r \neq \emptyset$ if and only if $k \in$ FIRST and $(k, \bot) \in F_r$. By the definition of F_r, this is equivalent to $k \in$ FIRST and $k \in$ LAST, which in turn holds if and only if $((x, y) \circ'_r a_k) \cap F \neq \emptyset$.

- If $(x, y) = (\bot, j)$, then we have $((x, y) \cdot'_r a_k) \cap F_r \neq \emptyset$ if and only if $k \in$ FIRST and $(k, j) \in F_r$. By definition of F_r, this is equivalent to $k \in$ FIRST and $j \in$ FOLLOW(k). But this again holds if and only if we have $(\bot, k) \in F_r$ and $j \in$ FOLLOW(k), that is, if and only if $((x, y) \circ'_r a_k) \cap F_r \neq \emptyset$.

- If $(x, y) = (i, \bot)$, then $((x, y) \cdot'_r a_k) \cap F_r \neq \emptyset$ if and only if $k \in$ FOLLOW(i) and $(k, \bot) \in F_r$. By the definition of F_r, this holds if and only if k belongs to both sets FOLLOW(i) and LAST. But this again holds if and only if $(i, k) \in F_r$ and $k \in$ LAST, that is, if and only if $((x, y) \circ'_r a_k) \cap F_r \neq \emptyset$.

- Finally, if $(x, y) = (i, j)$, then we have $((x, y) \cdot'_r a_k) \cap F_r \neq \emptyset$ if and only if $k \in$ FOLLOW(i) and $(k, j) \in F_r$. By the definition of F_r, this holds if and only if $k \in$ FOLLOW(i) and $j \in$ FOLLOW(k). But this again holds if and only if $(i, k) \in F_r$ and $j \in$ FOLLOW(k), that is, if and only if $((x, y) \circ'_r a_k) \cap F_r \neq \emptyset$.

This proves that A'_r has the F-property.

Now the proof of $L(A'_r) = L(r')$ is easy, since we know from Corollary 6.11 that A'_r accepts a word $w \in \Gamma^*$ if and only if $((\bot, \bot) \cdot'_r w) \cap F_r \neq \emptyset$. Since the forward transitions of A'_r exactly simulate the behaviour of the classical Glushkov automaton [40] for the language $L(r')$, we obtain $L(A'_r) = L(r')$. Finally we come to the automaton A_r. From the definition of $\cdot_r$ and $\circ_r$ one can see that the F-property of A'_r implies the F-property for A_r. Further, also the $\diamond$-property translates from A'_r to A_r because for all $(x, y) \in Q_r$ and $a, b \in \Sigma$ we have

$$((x,y) \cdot_r a) \circ_r b = \bigcup_{\substack{\varphi(a_i)=a \\ \varphi(b_i)=b}} ((x,y) \cdot'_r a_i) \circ'_r b_i = \bigcup_{\substack{\varphi(a_i)=a \\ \varphi(b_i)=b}} ((x,y) \circ'_r b_i) \cdot'_r a_i = ((x,y) \circ_r b) \cdot_r a.$$

Since $\varphi(L(r')) = L(r)$ and $L(A_r) = \varphi(L(A'_r))$, we obtain $L(A_r) = L(r)$. □

The following result provides a lower bound for this conversion.

Theorem 8.7. *For all integers $n \geq 1$ there is a binary language L_n with alphabetic width n, such that any nondeterministic biautomaton with the $\diamond$- and the F-property needs n^2 states to accept the language L_n.*

Proof. Let $n \geq 1$ and consider the language L_n which is described by the regular expression $r = (a^{n-1}b)^*$ of alphabetic width n. We show that the set

$$S = \{\, (a^i, a^{n-1-i}ba^{n-1-j}, a^j b) \mid 0 \leq i, j \leq n-1 \,\}$$

is a bi-fooling set for the language L_n, which by Lemma 8.4 proves the statement. First note that we have $a^i \cdot a^{n-1-i}ba^{n-1-j} \cdot a^j b = (a^{n-1}b)^2 \in L_n$, for all i

and j with $0 \leq i, j \leq n - 1$. Now consider two distinct pairs $(i, j) \neq (i', j')$, with $0 \leq i, i', j, j' \leq n - 1$. If $i \neq i'$, then we may assume that $i < i'$. Then the word $a^i \cdot a^{n-1-i'} b a^{n-1-j'} \cdot a^j b$ does not belong to L_n, since the number of a symbols before the first b symbol is less than $n - 1$. If otherwise $i = i'$, then it must be $j \neq j'$, and we may assume that $j < j'$. In this case the word $a^i \cdot a^{n-1-i'} b a^{n-1-j'} \cdot a^j b$ does not belong to L_n, because the number of a symbols between the first and the second b symbol is less than $n - 1$. Thus, the set S is a bi-fooling set for L_n. Since S is of size n^2, the statement of the theorem follows. $\qquad\square$

What can we say about the cost of converting a regular expression into a *deterministic* biautomaton with $\diamond$- and F-property? From the polynomial size conversion of regular expressions to $\diamond$-F-NBiAs, given in Theorem 8.6, together with the exponential upper bound of determinization, see Corollary 8.1, we obtain an exponential upper bound. Further, an exponential lower bound follows from the exponential cost of converting regular expressions into deterministic finite automata: it was shown by Hing Leung [82] that the minimal DFA for the language described by the regular expression $r_n = (a + (ab^*)^{n-1} a)^*$ needs 2^n states. Since any $\diamond$-F-DBiA contains an equivalent DFA, see Fact 6.13, and the alphabetic width of r_n is linear in n, we obtain an exponential blow-up in the description size when converting regular expressions to $\diamond$-F-DBiAs. The search for an asymptotically tight bound for the cost of this conversion is left for further research.

8.3   From Syntactic Monoids to Biautomata

Every language L over an alphabet Σ induces an equivalence relation $\equiv_L$ on Σ^*, called the *syntactic congruence* of L—we recall its definition from Section 2.1: two words u and v from Σ^* satisfy $u \equiv_L v$ if and only if for all $x, y \in \Sigma^*$ we have $xuy \in L$ if and only if $xvy \in L$. The set $M(L) = \Sigma^* / \equiv_L$ of equivalence classes of $\equiv_L$ is the *syntactic monoid* of L. It is known that L is a regular language if and only if the syntactic monoid $M(L)$ is finite [98, 105]. Notice that different languages may have the same underlying syntactic monoid. For example, one can see from the definition of $\equiv_L$ that the syntactic monoid $M(L)$ of a language $L \subseteq \Sigma^*$ coincides with the syntactic monoid $M(\overline{L})$ of the complement language $\overline{L} = \Sigma^* \setminus L$. Nevertheless, the language L can be described by its syntactic monoid $M(L)$ and a subset $N \subseteq M(L)$, consisting of those equivalence classes of $\equiv_L$ the union of which gives L. If we denote by $\varphi \colon \Sigma^* \to M(L)$ the morphism that maps every word $w \in \Sigma^*$ to its equivalence class $[w]_{\equiv_L}$, then we have $L = \varphi^{-1}(N)$.

Jirásková and Klíma [68] showed that if a regular language L is given by its syntactic monoid of size n, then the minimal deterministic biautomaton with $\diamond$- and F-property for L has at most n^2 states and this bound can be reached for certain n. In the following we show that we can do better in case of nondeterministic biautomata.

Theorem 8.8. *Let L be regular language given by a syntactic monoid of size n. Then the minimal nondeterministic biautomaton with the $\diamond$- and the F-property has at most n states. This bound can be reached by unary languages for every integer $n \geq 1$.*

Proof. Let $M(L)$ be the syntactic monoid of $L \subseteq \Sigma^*$. We refer to the monoid operation by $\bullet$. Moreover, let $\varphi : \Sigma^* \to M(L)$ be the morphism and choose a subset $N \subseteq M(L)$, such that $L = \varphi^{-1}(N)$, that is, $N = \{\, \varphi(u) \mid u \in L \,\}$. Then we define a nondeterministic biautomaton A_φ with $L(A_\varphi) = L$ as follows: let $A_\varphi = (Q, \Sigma, \cdot, \circ, N, F)$, with states $Q = M(L)$, final states $F = \{1\}$— here $1 = \varphi(\lambda)$ is the identity element of $M(L)$—and

$$m \cdot a = \{\, n \in M(L) \mid m = \varphi(a) \bullet n \,\}, \quad m \circ a = \{\, n \in M(L) \mid m = n \bullet \varphi(a) \,\},$$

for every $m \in M(L)$ and $a \in \Sigma$. Before we prove that biautomaton A_φ accepts the language L, we first show that A_φ satisfies both the $\diamond$-property and the F-property. For the $\diamond$-property we argue as follows:

$$
\begin{aligned}
(m \cdot a) \circ b &= \{\, n \in M(L) \mid m = \varphi(a) \bullet n \,\} \circ b \\
&= \bigcup_{n :\, m = \varphi(a) \bullet n} \{\, r \in M(L) \mid n = r \bullet \varphi(b) \,\} \\
&= \{\, r \in M(L) \mid m = \varphi(a) \bullet (r \bullet \varphi(b)) \,\} \\
&= \{\, r \in M(L) \mid m = (\varphi(a) \bullet r) \bullet \varphi(b) \,\} \\
&= \bigcup_{n :\, m = n \bullet \varphi(b)} \{\, r \in M(L) \mid n = \varphi(a) \bullet r \,\} \\
&= \{\, n \in M(L) \mid m = n \bullet \varphi(b) \,\} \cdot a \\
&= (m \circ b) \cdot a.
\end{aligned}
$$

Second, the F-property of A_φ is seen as follows: observe, that $[m \cdot a] \cap F \neq \emptyset$ if and only if $1 \in \{\, n \in M(L) \mid m = \varphi(a) \bullet n \,\}$, that is, if and only if $m = \varphi(a) \bullet 1$. This in turn is equivalent to $m = 1 \bullet \varphi(a)$, because 1 is the identify element of the monoid $M(L)$. Therefore we obtain $1 \in \{\, n \in M(L) \mid m = n \bullet \varphi(a) \,\}$, which is equivalent to $[m \circ a] \cap F \neq \emptyset$. We have shown that for every element $m \in M(L)$ and symbol $a \in \Sigma$ the following holds:

$$[m \cdot a] \cap F \neq \emptyset \quad \text{if and only if} \quad [m \circ a] \cap F \neq \emptyset.$$

Now we turn to verify $L = L(A_\varphi)$. Because A_φ has the $\diamond$- and the F-property, we know from Corollary 6.11 that a word $w \in \Sigma^*$ is accepted by A_φ if and only if $(N \cdot w) \cap F \neq \emptyset$—recall that N is the set of initial states of A_φ. In our case this means that w is accepted if and only if $1 \in N \cdot w$. By induction on the length of w one can see that

$$N \cdot w = \{\, r \in M(L) \mid \varphi(w) \bullet r \in N \,\},$$

so w is accepted by A_φ if and only if $\varphi(w) \bullet 1 = \varphi(w) \in N$. Because $L = \varphi^{-1}(N)$ we obtain $w \in L(A_\varphi)$ if and only if $w \in L$.

It remains to prove that the bound n on the size of nondeterministic biautomata can be reached. To this end consider the unary language $L_n = (a^n)^*$, which is accepted by the syntactic monoid $M = \{ a^i \mid 0 \leq i < n \}$ with the operation $\bullet$ defined by $a^i \bullet a^j = a^{(i+j) \bmod n}$, for $0 \leq i, j \leq n - 1$, the trivial morphism $\varphi(a) = a$, and the set $N = \{a^0\}$. Clearly we have $L_n = \varphi^{-1}(N)$. Since any nondeterministic biautomaton with the $\diamond$- and the F-property that accepts L_n must "contain" a nondeterministic finite automaton accepting L_n and no nondeterministic finite automaton with less than n states can accept this language, the lower bound follows. $\qquad\square$

We summarize the descriptional complexity of converting between the before studied models of describing regular languages in Figure 8.4. The tight 2^n bound for converting nondeterministic to deterministic finite automata is well known, the bounds for the conversions of syntactic monoids and finite automata into $\diamond$-F-DBiAs were proven by Jiráskova and Klíma [68]. The bounds dealing with nondeterministic biautomata have been shown in the previous sections. Recall that exponential upper and lower bounds for the conversion of regular expressions into $\diamond$-F-DBiAs were discussed at the end of Section 8.2.

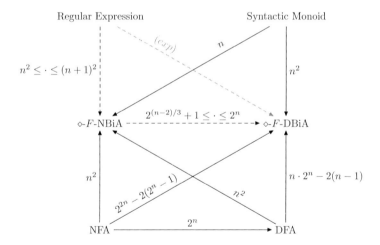

Figure 8.4: Descriptional complexity of conversion problems for biautomata with $\diamond$- and F-property. Solid arrows denote tight bounds for the corresponding conversions, and dashed arrows denote conversions where the exact bound lies between the presented lower and upper bound. The grayed out dashed arrow connecting regular expressions to $\diamond$-F-DBiAs indicates that exponential upper and lower bounds can be derived for this conversion.

8.4 Complexity of the $\diamond$- and F-properties

In the previous sections we only considered biautomata that have both the $\diamond$-property and the F-property. Naturally the question arises, how costly it is to convert a biautomaton that does not have both these properties into a biautomaton that does. We first consider the complexity of the F-property. We will see that this property can easily be enforced for nondeterministic biautomata or if we do not require to preserve the $\diamond$-property (if present). On the other hand, for deterministic biautomata with $\diamond$-property, the cost of constructing an equivalent deterministic biautomaton with both properties is exponential.

We start with biautomata which do not need to satisfy the $\diamond$-property. In this case we have seen in the proof of Lemma 7.9, how to convert a biautomaton into one that has the F-property: let $A = (Q, \Sigma, \cdot, \circ, I, F)$ be a biautomaton, and define the biautomaton $B = (Q_B, \Sigma, \cdot_B, \circ_B, I, F_B)$, with $Q_B = Q \cup \{q_f\}$, $F_B = F \cup \{q_f\}$, for a new accepting state $q_f \notin Q$, and the transitions $\cdot_B$ and $\circ_B$ such that for all states $q \in Q$ and input symbols $a \in \Sigma$ we have

$$
q \cdot_B a = \begin{cases} (q \cdot a) \cup \{q_f\} & \text{if } (q \cdot a) \cap F \neq \emptyset \text{ or } (q \circ a) \cap F \neq \emptyset, \\ q \cdot a & \text{otherwise,} \end{cases}
$$

$$
q \circ_B a = \begin{cases} (q \circ a) \cup \{q_f\} & \text{if } (q \cdot a) \cap F \neq \emptyset \text{ or } (q \circ a) \cap F \neq \emptyset, \\ q \circ a & \text{otherwise.} \end{cases}
$$

For the new state q_f we have $q_f \cdot a = q_f \circ a = \emptyset$, for all symbols $a \in \Sigma$. It is shown in Lemma 7.9 that $L(A) = L(B)$. Therefore, nondeterministic biautomata may be forced to satisfy the F-property by the expense of a single additional state. We summarize this in the following lemma.

Lemma 8.9. *For any given nondeterministic biautomaton with n states, one can construct an equivalent nondeterministic biautomaton with $n+1$ states that satisfies the F-property.* $\square$

For deterministic biautomata, we derive a linear bound for this conversion as follows: assume the biautomaton A is deterministic, and construct the nondeterministic biautomaton B as described above. Then the corresponding power-set automaton $\mathcal{P}(B)$ has at most twice the number of states as A, namely the singleton states $\{q\}$, for $q \in Q$, and states of the form $\{q, q_f\}$, for $q \in Q$. This gives us the following result.

Corollary 8.10. *For any given deterministic biautomaton with n states, one can construct an equivalent deterministic biautomaton with $2n$ states that satisfies the F-property.* $\square$

Next we consider nondeterministic biautomata that satisfy the $\diamond$-property. The following lemma provides a polynomial upper bound for the conversion of $\diamond$-NBiAs to $\diamond$-F-NBiAs.

Lemma 8.11. *For any given nondeterministic biautomaton with n states that has the $\diamond$-property, one can construct an equivalent nondeterministic biautomaton with $O(n^4)$ states, that has both the $\diamond$-property and the F-property.*

Proof. Let $A = (Q, \Sigma, \cdot, \circ, I, F)$ be a biautomaton that has the $\diamond$-property. We have seen in the proof of Theorem 7.11 that

$$L(A) = \bigcup_{q \in Q} \{\, u \mid q \in I \cdot u \,\} \cdot \{\, v \mid (q \circ v) \cap F \neq \emptyset \,\}$$

holds, and that for every state $q \in Q$ the two languages $L_1(q) = \{\, u \mid q \in I \cdot u \,\}$ and $L_2(q) = \{\, v \mid (q \circ v) \cap F \neq \emptyset \,\}$ can be accepted by nondeterministic finite automata, each having n states. By standard methods we construct, for all $q \in Q$, a nondeterministic finite automaton with $2n$ states that accepts $L_1(q) \cdot L_2(q)$. From these n different automata we construct a nondeterministic finite automaton B, with $L(B) = L(A)$, that has $O(n^2)$ states. Finally, by Theorem 8.3, this finite automaton can be transformed into a $\diamond$-F-NBiA C, with $L(C) = L(B)$, having $O(n^4)$ states. $\qquad\square$

It remains to consider deterministic biautomata with the $\diamond$-property that do not satisfy the F-property. The next theorem shows that, from a descriptional complexity point of view, it is expensive to transform such a biautomaton into a biautomaton that has both the $\diamond$-property and the F-property.

Theorem 8.12. *Let $n \geq 1$. There is a deterministic biautomaton A_n that satisfies the $\diamond$-property and that has $(n+2)^2 + 1$ states, such that any deterministic biautomaton that accepts $L(A_n)$ and that satisfies both the $\diamond$-property and the F-property needs at least $\Omega(2^n)$ states.*

Proof. Consider the regular language $L_n = (a+b)^n \cdot a \cdot (a+b)^* \cdot a \cdot (a+b)^n$. It is easy to see that every deterministic finite automaton accepting L_n needs $\Omega(2^n)$ states. Since the minimal deterministic finite automaton is contained in the minimal deterministic biautomaton with both the $\diamond$-property and the F-property, also every such biautomaton needs $\Omega(2^n)$ states. But if we do not require the F-property, the language L_n can also be accepted by a deterministic biautomaton A_n with the $\diamond$-property with $(n+2)^2 + 1$ states. The states of this biautomaton are pairs (i,j), with $i,j \in \{0, 1, \ldots, n, f\}$, and additionally a sink state s. The initial state is $(0,0)$, the only accepting state is (f,f), and the transition functions $\cdot$ and $\circ$ are defined by

$$(i,j) \cdot a = \begin{cases} (i+1, j) & \text{if } i \notin \{n, f\}, \\ (f, j) & \text{if } i \in \{n, f\}, \end{cases} \qquad (i,j) \circ a = \begin{cases} (i, j+1) & \text{if } j \notin \{n, f\}, \\ (i, f) & \text{if } j \in \{n, f\}, \end{cases}$$

$$(i,j) \cdot b = \begin{cases} (i+1, j) & \text{if } i \notin \{n, f\}, \\ s & \text{if } i = n, \\ (f, j) & \text{if } i = f, \end{cases} \qquad (i,j) \circ b = \begin{cases} (i, j+1) & \text{if } j \notin \{n, f\}, \\ s & \text{if } j = n, \\ (i, f) & \text{if } j = f, \end{cases}$$

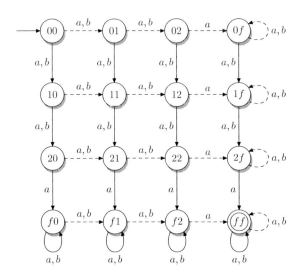

Figure 8.5: The deterministic biautomaton A_n, for $n = 2$, that accepts the regular language $L_n = (a + b)^2 \cdot a \cdot (a + b)^* \cdot a \cdot (a + b)^2$. The non-accepting sink state s and all transitions leading to it are not shown. This biautomaton satisfies the $\diamond$-property, but not the F-property.

for every $i, j \in \{0, 1, \ldots, n, f\}$. The sink state s goes to itself on every transition. Figure 8.5 shows the automaton A_n for $n = 2$. The biautomaton A_n counts the number of symbols it has read from the left in the first component of the state and the number of symbols it has read from the right in the second component of the state. If a counter reaches n, then the next symbol in the corresponding reading direction must be an a. Because the transition function $\cdot$ only operates on the first component of a state and the function $\circ$ only operates on the second component of a state, one can see that A_n has the $\diamond$-property. This concludes our proof. $\qquad\square$

Using the approach of Lemma 8.11 and a subsequent power-set construction, one can also deduce an exponential upper bound for the cost of converting $\diamond$-DBiAs to $\diamond$-F-DBiAs. The search for a tight bound is left for further research.

It remains to study the problem of converting a biautomaton that does not have the $\diamond$-property into one that has this property. Obviously, this is not always possible, because we have seen in Chapter 7 that biautomata in general accept linear context-free languages, while biautomata that have the $\diamond$-property can accept only regular languages. Nevertheless, we can consider the class of all biautomata that accept only regular languages, and ask for the cost of converting such biautomata into biautomata that obey the $\diamond$-property. It turns out that the trade-off between these two descriptive models is non-recursive.

A *non-recursive trade-off* from a descriptive model $\mathcal{A}$ to a descriptive model $\mathcal{B}$ means that there is *no* recursive function that serves as an upper bound for the increase of the description size, when changing the representation of a language from model $\mathcal{A}$ to model $\mathcal{B}$. More general studies of non-recursive trade-offs can be found, for example, in papers of Gruber et al. [43] and Kutrib [79].

An early result from Meyer and Fischer [94] shows that the trade-off between context-free grammars and finite automata is non-recursive. In fact, it is known that also the trade-off between linear context-free languages and finite automata is non-recursive, see for example the papers of Kutrib [79] and Malcher [87]. We can use this to deduce a non-recursive trade-off between biautomata in general and biautomata that satisfy the ◇-property as follows. Recall that the representation of a language by a linear context-free grammar can effectively be converted first into a linear context-free grammar in normal form, and then, by Theorem 7.6, into a biautomaton. Therefore, the trade-off between linear context-free grammars and biautomata *is recursive*. Further, because of the effective conversion of ◇-NBiAs to ◇-F-NBiAs from Lemma 8.11, and the fact that any ◇-F-NBiA contains an equivalent nondeterministic finite automaton, we also get a recursive trade-off between biautomata with the ◇-property, and finite automata. Now, if the trade-off between biautomata in general and biautomata with the ◇-property would be recursive, then we also get a recursive trade-off between linear context-free grammars and finite automata. Since the latter trade-off is non-recursive, the trade-off between biautomata in general and such with the ◇-property is non-recursive, too. We have shown the following result.

Theorem 8.13. *The trade-off between deterministic or nondeterministic biautomata and deterministic or nondeterministic biautomata that satisfy the ◇-property is non-recursive.* □

Chapter 9

Minimization Algorithms for Biautomata

IN this chapter we discuss the minimization of biautomata. First we will show that deterministic biautomata satisfying both the $\diamond$-property and the F-property can be minimized with similar methods as ordinary deterministic finite automata, even with a Brzozowski-like algorithm—see Section 9.2. On the other hand, we will see in Section 9.3 that minimizing biautomata that do not have both these properties becomes more involved.

9.1 Efficient State Merging Algorithms

We have seen in Section 6.2.1 that deterministic biautomata with the $\diamond$-property and the F-property show a lot of similarities to classical deterministic finite automata. In the following we show that classical minimization methods for deterministic finite automata can also be used to minimize deterministic biautomata with $\diamond$-property and F-property. The problem of minimizing biautomata that lack one or both of these properties will be studied in Section 9.3.

Let us first recall some notions that were used by Klíma and Polák [74]. Let $A = (Q, \Sigma, \cdot, \circ, q_0, F)$ be a deterministic biautomaton with $\diamond$- and F-property. A relation $\simeq\ \subseteq Q \times Q$ is a *congruence* of A if

- $\simeq$ is an equivalence relation on Q,

- for all states $p, q \in Q$ and letters $a \in \Sigma$, if $p \simeq q$, then $(p \cdot a) \simeq (q \cdot a)$ and $(p \circ a) \simeq (q \circ a)$,

- for all states $p \in F$ and $q \in Q$, if $p \simeq q$, then $q \in F$.

The equivalence class of a state $q \in Q$ under the congruence $\simeq$ is defined as $[q]_\simeq = \{\, p \in Q \mid p \simeq q \,\}$, and the set of all equivalence classes of Q under $\simeq$ is denoted by $Q/\!\simeq\ = \{\, [q]_\simeq \mid q \in Q \,\}$. Further, let $F/\!\simeq\ = \{\, [q]_\simeq \mid q \in F \,\}$.

Now the *factor biautomaton* of A under the congruence $\simeq$ is the biautomaton $A_\simeq = (Q/\simeq, \Sigma, \cdot_\simeq, \circ_\simeq, [q_0]_\simeq, F/\simeq)$, where the transition functions are such that $[q]_\simeq \cdot_\simeq a = [q \cdot a]_\simeq$ and $[q]_\simeq \circ_\simeq a = [q \circ a]_\simeq$, for all states $q \in Q$ and letters $a \in \Sigma$. Note that the functions $\cdot_\simeq$ and $\circ_\simeq$ are well defined because $\simeq$ is a congruence. It is known that the automaton $A_\simeq$ is also a deterministic biautomaton, with $\diamond$-property and F-property, that satisfies $L(A_\simeq) = L(A)$ [74].

Notice that *equivalence of states* in a deterministic biautomaton A with $\diamond$- and F-property, that is, the relation $\equiv$ defined in Section 6.1 as $p \equiv q$ if and only if $L_A(p) = L_A(q)$, is a congruence relation on A. Therefore, combining all states that are in the same equivalence class under the relation $\equiv$ into a single state, results in a deterministic biautomaton $A_\equiv$, which still has the $\diamond$- and F-properties, and which satisfies $L(A_\equiv) = L(A)$. This combining of states can be done by state merging, which is defined similar as for finite automata—see Section 2.3: consider a deterministic biautomaton $A = (Q, \Sigma, \cdot, \circ, q_0, F)$ and two of its states p and q. The biautomaton that is obtained by *merging* state p to state q is the automaton $A' = (Q \setminus \{p\}, \Sigma, \cdot', \circ', q_0', F \setminus \{p\})$, where $q_0' = q$, if $q_0 = p$, and otherwise, if $q_0 \neq p$, then $q_0' = q_0$. The transition functions $\cdot'$ and $\circ'$ are defined by

$$ r \cdot' a = \begin{cases} q, & \text{if } r \cdot a = p, \\ r \cdot a & \text{otherwise}, \end{cases} \quad \text{and} \quad r \circ' a = \begin{cases} q, & \text{if } r \circ a = p, \\ r \circ a & \text{otherwise}. \end{cases} $$

Assume we want to construct the factor biautomaton $A_\equiv$ by merging equivalent states. If we do a stepwise state merging, instead of directly combining all equivalent states at once, then we may construct intermediate biautomata, which do *not* have the $\diamond$-property. This is illustrated in the following example.

Example 9.1. Consider the deterministic biautomaton A that is shown on the left-hand side in Figure 9.1. Notice that this automaton has both the $\diamond$-property and the F-property. Further, the states q_1 and q_2 are equivalent, because their right languages $L_A(q_1) = \{a\} \cdot \{aa\}^*$ and $L_A(q_2) = \{a\} \cdot \{aa\}^*$ are equal.

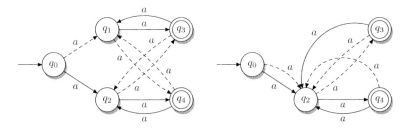

Figure 9.1: Left: a deterministic biautomaton A with $\diamond$- and F-property, where merging any state to an equivalent state destroys the $\diamond$-property. Right: the deterministic biautomaton A' with the F-property but without the $\diamond$-property, that results from A by merging state q_1 to state q_2.

Also the states q_3 and q_4 are equivalent, because their right languages are $L_A(q_3) = \{aa\}^* = L_A(q_4)$. If we merge state q_1 to state q_2, then the resulting biautomaton A', which is depicted on the right-hand side in Figure 9.1, does not satisfy the $\diamond$-property: let $\cdot'$ and $\circ'$ be the forward and backward transition functions of the biautomaton A', then

$$(q_0 \cdot' a) \circ' a = q_2 \circ' a = q_3 \neq q_4 = q_2 \cdot' a = (q_0 \circ' a) \cdot' a.$$

Similarly, if we merge state q_2 to state q_1, then the resulting biautomaton does not satisfy the $\diamond$-property either. The same happens if we merge state q_3 to state q_4 instead, or if we merge state q_4 to state q_3. Assume for example that state q_3 is merged to state q_4 and let $\cdot''$ and $\circ''$ be the resulting forward and backward transition functions, respectively. Then

$$(q_1 \cdot'' a) \circ'' a = q_4 \circ'' a = q_1 \neq q_2 = q_4 \cdot'' a = (q_1 \circ'' a) \cdot'' a.$$

So in the automaton A we cannot merge any single pair of equivalent states without loosing the $\diamond$-property.

Nevertheless, if we also merge in the automaton A' state q_3 to its equivalent state q_4, then the resulting biautomaton $\tilde{A}$, which is depicted in Figure 9.2, *does* satisfy the $\diamond$-property again. Notice that no two states of the biautomaton $\tilde{A}$ are equivalent. We will see later that this fact implies that the deterministic biautomaton $\tilde{A}$, which has both the $\diamond$- and the F-property, is minimal. $\diamond$

If all states of a deterministic biautomaton A that satisfies the $\diamond$-property and the F-property are reachable, then the biautomaton $A_\equiv$ turns out the be a *minimal* deterministic biautomaton with $\diamond$- and F-property for the language $L(A)$. This can be concluded by the upcoming Lemma 9.2 of Klíma and Polák [74]. Even more, the following lemma also shows that all minimal deterministic biautomata with $\diamond$- and F-property for a regular language L are isomorphic in the following sense—see [74]: two biautomata $A = (Q, \Sigma, \cdot, \circ, q_0, F)$ and $A' = (Q', \Sigma, \cdot', \circ', q_0', F')$ are called *isomorphic* if there is a bijection $h \colon Q \to Q'$, called an *isomorphism*, such that the following conditions hold:

- $h(q_0) = q_0'$,

- for all $q \in Q$ and $a \in \Sigma$, we have $h(q \cdot a) = h(q) \cdot' a$ and $h(q \circ a) = h(q) \circ' a$,

- for all $q \in Q$, we have $q \in F$ if and only if $h(q) \in F'$.

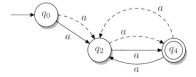

Figure 9.2: The deterministic biautomaton $\tilde{A}$ with $\diamond$- and F-property that results from A' by merging state q_3 to state q_4.

Now the result of Klíma and Polák [74] reads as follows.

Lemma 9.2. *Let $A = (Q, \Sigma, \cdot, \circ, q_0, F)$ be deterministic biautomaton that has both the $\diamond$- and the F-property, and where all states are reachable. Let $L = L(A)$. Then the relation $\equiv$ is a congruence relation on A. Moreover the mapping*

$$h \colon [(q_0 \cdot u) \circ v]_{\equiv} \mapsto u^{-1} L v^{-1}$$

is an isomorphism between the factor biautomaton $A_{\equiv}$ and the canonical biautomaton A_L for the language $L = L(A)$.

It follows directly from Lemma 9.2 that the factor biautomaton $A_{\equiv}$ is minimal: assume to the contrary that there is another deterministic biautomaton B, with $L(B) = L(A_{\equiv})$, that has the $\diamond$- and the F-property, and that has fewer states than $A_{\equiv}$. Then Lemma 9.2 implies that both automata $B_{\equiv}$ and $A_{\equiv}$ are isomorphic to the canonical biautomaton of the language $L(B) = L(A_{\equiv})$. But then $B_{\equiv}$ and $A_{\equiv}$ have the same number of states and therefore, B cannot have fewer states than $A_{\equiv}$. From above considerations we can conclude the following characterization of minimal deterministic biautomata with $\diamond$- and F-property.

Theorem 9.3. *A deterministic biautomaton A that has both the $\diamond$-property and the F-property is minimal if and only if all states of A are reachable and no two distinct states of A are equivalent.*

Proof. Clearly, a minimal biautomaton cannot have any unreachable states. Let $A = (Q, \Sigma, \cdot, \circ, q_0, F)$ be a biautomaton with the $\diamond$-property and the F-property, such that all of its states are reachable. Then the factor biautomaton $A_{\equiv}$ is a biautomaton with $\diamond$- and F-property, that is equivalent to A. If there are distinct states p and q, with $p \equiv q$, then the automaton $A_{\equiv}$ has fewer states than A. In this case A cannot be minimal. Now assume that no two distinct state of A are equivalent. Then both biautomata A and $A_{\equiv}$ have the same number of states. Since $A_{\equiv}$ is minimal, also A is a minimal automaton. □

By Corollary 6.11 we know that if A is a biautomaton that has both the $\diamond$-property and the F-property, then the right language $L_A(q)$ of a state q in A is $L_A(q) = \{ w \in \Sigma^* \mid q_0 \cdot w \in F \}$. So this language only depends on the forward transitions of A. Therefore, deciding whether two states p and q of such a biautomaton are equivalent can be done as follows: treat the automaton like a classical deterministic finite automaton by ignoring all backward transitions and check whether the states p and q are equivalent in this finite automaton.

We can also compute all pairs of states in a deterministic biautomaton with $\diamond$- and F-property that are *not* equivalent, by using a similar approach as for finite automata. First, we mark as *distinguishable* all pairs of states (p, q), where $p \in F$ and $q \notin F$, or where $p \notin F$ and $q \in F$. Then, as long as there is an unmarked pair (p, q) and a letter $a \in \Sigma$, such that the pair $(p \cdot a, q \cdot a)$ or the pair $(p \circ a, q \circ a)$ is marked, we also mark pair (p, q) as distinguishable. When no new marking can be done, we have marked all pairs of distinguishable states in the automaton, so two states p and q are equivalent if and only if the pair (p, q) is not marked. Again, Corollary 6.11 allows us to concentrate only on forward transitions, so in the above algorithm it remains to look for unmarked pairs (p, q) and letters $a \in \Sigma$, such that the pair $(p \cdot a, q \cdot a)$ is marked.

Now the minimization of deterministic biautomata with $\diamond$- and F-property can be done very similar to minimization methods for deterministic finite automata. In a first step, we remove all unreachable states. Now we ignore all backward transitions, treating the automaton like a deterministic finite automaton, and calculate state equivalence in the automaton. Afterwards we perform the merging of equivalent states in the biautomaton. In the following Section 9.2 we will see that Brzozowski's minimization algorithm for finite automata, which uses the power-set and reversal constructions twice to compute the minimal deterministic finite automaton, also works for biautomata.

9.2 Brzozowski-Like Minimization of Biautomata

In Section 9.1 we have shown how to adapt classical minimization algorithms for deterministic finite automata to the case of deterministic biautomata with the $\diamond$- and F-property. Another interesting algorithm for minimizing deterministic finite automata is Brzozowski's finite automaton minimization algorithm [12]: given a (deterministic or nondeterministic) finite automaton A, it computes the automaton $\mathcal{P}([\mathcal{P}(A^R)]^R)$, which turns out to be the minimal deterministic finite automaton for $L(A)$. Adaptations of this algorithm for lossy compression of finite automata have been presented in Chapter 5. While the minimality of the constructed automaton $\mathcal{P}([\mathcal{P}(A^R)]^R)$ needs some argumentation, the facts that it accepts the correct language and that it is a deterministic finite automaton are easy to see, since the dual B^R of a finite automaton B accepts the reverse of the language accepted by B, that is, $L(B^R) = L(B)^R$.

We have seen the Section 7.3 that the relation between the languages accepted by a biautomaton A and its dual A^R is not as simple as for classical finite automata. Nevertheless, we can show that Brzozowski's minimization algorithm can still be used for minimization of biautomata. More precisely, we prove that for every (deterministic or nondeterministic) biautomaton A, with both the $\diamond$- and the F-property, the automaton $\mathcal{P}([\mathcal{P}(A^R)]^R)$ is the unique minimal deterministic biautomaton that has both the $\diamond$- and the F-property. Of course, the power-set construction as defined in the beginning of Chapter 7 may produce unreachable states, which must not be present in a minimal automaton. Therefore, in this chapter we assume that $\mathcal{P}(A)$ contains only those states which are reachable from its initial state.

Remember that by Lemma 7.18, the middle automaton $\mathcal{P}(A^R)$ in the computation of $\mathcal{P}([\mathcal{P}(A^R)]^R)$, for some biautomaton A with $\diamond$- and F-property, is a deterministic biautomaton that has the $\diamond$- and the I-property. The following lemma starts with this middle automaton.

Lemma 9.4. *Let A be a deterministic biautomaton with both the $\diamond$- and the I-property and with no unreachable states. Then $\mathcal{P}(A^R)$ is a minimal biautomaton with both the $\diamond$- and the F-property.*

Proof. Let $A = (Q, \Sigma, \cdot, \circ, q_0, F)$ be a deterministic biautomaton that satisfies both the $\diamond$- and the I-property and where all states are reachable. Further, let $A^R = (Q, \Sigma, \cdot^R, \circ^R, F, \{q_0\})$ be the dual of A, and $B = (Q_B, \Sigma, \cdot_B, \circ_B, q_0^B, F_B)$ the power-set biautomaton of A^R, that is, $B = \mathcal{P}(A^R)$. Lemmas 7.1 and 7.18 imply that B has the $\diamond$-property and the F-property. In the following, we prove that B does not have a pair of distinct but equivalent states. Since we assumed the power-set construction to build only the reachable part of the automaton, all states in $B = \mathcal{P}(A^R)$ are reachable and the minimality of B then follows from Theorem 9.3.

Let P_1 and P_2 be two distinct states of B, then we may assume that there is an element $q \in Q$, with $q \in P_1 \setminus P_2$. Since q is reachable in the deterministic biautomaton A, there are words $u, v \in \Sigma^*$, such that $q = (q_0 \cdot u) \circ v$. Since A has both the $\diamond$- and the I-property, Lemma 6.12 implies $q = q_0 \cdot vu$. This means that in the dual automaton A^R, we have $q_0 \in q \cdot^R (vu)^R$, by Lemma 7.16. Furthermore, this means that in biautomaton B we have $q_0 \in (P_1 \cdot_B (vu)^R)$ because $q \in P_1$. Since the accepting states of B are the sets $P \in Q_B$ with $q_0 \in P$, it follows that the word $(vu)^R$ is accepted by B when starting from state P_1. Now assume, for the sake of contradiction, that $(vu)^R$ is also accepted by B when starting from state P_2. Then, since B has the $\diamond$- and the F-property, we know from Corollary 6.11 that $q_0 \in P_2 \cdot_B (vu)^R$. This means that there is a state $p \in P_2$, with $q_0 \in p \cdot^R (vu)^R$. From Lemma 7.16 and the fact that A is deterministic, we obtain $p = q_0 \cdot vu = q$, which is a contradiction to $q \notin P_2$. We have shown that $(vu)^R$ is accepted by B when starting from state P_1, but not when starting from state P_2, so P_1 and P_2 cannot be equivalent. □

Now we are ready to prove the main result of this section.

Theorem 9.5. *Let A be a (deterministic or nondeterministic) biautomaton that satisfies the $\diamond$-property and the F-property. Then $\mathcal{P}([\mathcal{P}(A^R)]^R)$ is the unique minimal biautomaton with $\diamond$-property and F-property for the language $L(A)$.*

Proof. Let $A = (Q, \Sigma, \cdot, \circ, I, F)$ be a biautomaton with both the $\diamond$-property and the F-property. Further, let $B = \mathcal{P}(A^R)$ and $C = \mathcal{P}(B^R)$. We denote these automata by $B = (Q_B, \Sigma, \cdot_B, \circ_B, q_0^B, F_B)$ and $C = (Q_C, \Sigma, \cdot_C, \circ_C, q_0^C, F_C)$, respectively. Then B is a deterministic biautomaton with the $\diamond$- and the I-property, with no unreachable states. So by Lemma 9.4, the automaton C is a minimal biautomaton with the $\diamond$- and the F-property. By Lemma 9.2, it follows that C is the unique minimal automaton among all biautomata with these two properties that accept $L(C)$. It remains to prove $L(A) = L(C)$, which, thanks to Corollary 6.11, can be done by only reasoning about forward transitions. For all words $w \in \Sigma^*$, we have $w \in L(A)$ if and only if $(I \cdot w) \cap F \neq \emptyset$. By Lemma 7.16, this holds if and only if $(F \cdot^R w^R) \cap I \neq \emptyset$, which is the same as $q_0^B \cdot_B w^R \in F_B$,

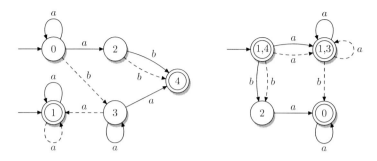

Figure 9.3: Left: The nondeterministic biautomaton A with the $\diamond$- and the F-property. Right: The deterministic biautomaton $B = \mathcal{P}(A^R)$ with the $\diamond$- and the I-property.

by definition of B. We again use Lemma 7.16, to see that $q_0^B \cdot_B w^R \in F_B$ holds if and only if $q_0^B \in F_B \cdot_B^R w$, which by definition of C is equivalent to $q_0^C \cdot_C w \in F_C$. Thus, we have $w \in L(A)$ if and only if $w \in L(C)$. $\qquad\square$

We illustrate the algorithm in the following example.

Example 9.6. Let $A = (Q, \{a, b\}, \cdot, \circ, I, F)$ be a nondeterministic biautomaton with $Q = \{0, 1, 2, 3, 4\}$, $I = \{0, 1\}$, $F = \{1, 4\}$, and with the transition functions $\cdot$ and $\circ$ as depicted on the left in Figure 9.3. Notice that A has both the $\diamond$-property and the F-property. The corresponding power-set biautomaton $B = \mathcal{P}(A^R)$ of the dual of A is depicted on the right in Figure 9.3. The sink state $\emptyset$ and all transitions leading to it are omitted. Finally, the minimal deterministic biautomaton $C = \mathcal{P}(B^R)$ is depicted in Figure 9.4, where again the sink state and corresponding transitions are omitted.

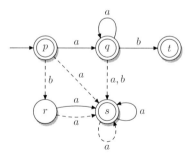

Figure 9.4: The minimal deterministic biautomaton $C = \mathcal{P}(B^R)$ with the $\diamond$- and F-property. The state symbols p, q, r, s, and t are abbreviations for subsets of the state set of B. We have $p = \{\{0\}, \{1, 3\}, \{1, 4\}\}$, $q = \{\{0\}, \{2\}, \{1, 3\}, \{1, 4\}\}$, $r = \{\{1, 3\}\}$, $s = \{\{1, 3\}, \{1, 4\}\}$, and $t = \{\{1, 4\}\}$.

Let us follow the argumentation in the proof of Lemma 9.4 to show that the states p and s cannot be equivalent. Notice that states $p = \{\, \{0\}, \{1,3\}, \{1,4\}\,\}$ and $s = \{\, \{1,3\}, \{1,4\}\,\}$ differ in the element $\{0\}$, which is reachable in the biautomaton B, for example by first reading a with a forward transition and then b with a backward transition: $(\{1,4\} \cdot a) \circ b = \{0\}$. One can check that by Lemma 6.12, this state of B is also reached by first reading b forwards, and then a forwards: $(\{1,4\} \cdot b) \cdot a = \{0\}$. This in turn means by Lemma 7.16 that the word ba is accepted from state p but not from state s, since $\{0\} \in p \setminus s$. Thus, states p and s cannot be equivalent. $\diamond$

We have seen that if we are given a deterministic or nondeterministic biautomaton A that has both the $\diamond$-property and the F-property, then the automaton $\mathcal{P}([\mathcal{P}(A^R)]^R)$ is the minimal deterministic biautomaton with $\diamond$- and F-property for the language $L(A)$. We will see in Section 9.3 that this is not always true for biautomata that do not have both properties, because the resulting automaton may not be minimal. Nevertheless, we can show that the automaton $\mathcal{P}([\mathcal{P}(A^R)]^R)$ at least accepts the same language as the biautomaton A, even if A does not have any of the structural properties.

Lemma 9.7. *Let A be a (deterministic or nondeterministic) biautomaton and let $C = \mathcal{P}([\mathcal{P}(A^R)]^R)$. Then $L(A) = L(C)$.*

Proof. Let $A = (Q_A, \Sigma, \cdot_A, \circ_A, I_A, F_A)$ be a biautomaton and A^R its dual biautomaton. Further, we denote by $B = (Q_B, \Sigma, \cdot_B, \circ_B, q_0^B, F_B)$ the power-set automaton of A^R, that is, $B = \mathcal{P}(A^R)$, and by B^R the dual biautomaton of B. Finally let $C = \mathcal{P}(B^R)$, denoted by $C = (Q_C, \Sigma, \cdot_C, \circ_C, q_0^C, F_C)$. Notice that the biautomata A and C are related by $C = \mathcal{P}([\mathcal{P}(A^R)]^R)$. We will show that $w \in L(A)$ if and only if $w \in L(C)$, for all words $w \in \Sigma^*$. By definition, the word w is accepted by biautomaton A if and only if there is a state $p_A \in I_A$, a state $q_A \in F_A$, and words $u_1, v_1, u_2, v_2, \ldots, u_k, v_k \in \Sigma^*$, with $w = u_1 u_2 \ldots u_k v_k \ldots v_2 v_1$, such that

$$q_A \in ((\ldots (((p_A \cdot_A u_1) \circ_A v_1) \cdot_A u_2) \circ_A v_2) \ldots) \cdot_A u_k) \circ_A v_k.$$

By Lemma 7.17 this is equivalent to

$$p_A \in ((((\ldots ((q_A \circ_A^R v_k^R) \cdot_A^R u_k^R) \ldots) \circ_A^R v_2^R) \cdot_A^R u_2^R) \circ_A^R v_1^R) \cdot_A^R u_1^R, \qquad (9.1)$$

which is a computation in the automaton A^R. Remember that the power-set biautomaton $B = \mathcal{P}(A^R)$ has the initial state $q_0^B = F_A$, and that the final states of B are subsets of Q_A that contain an element from I_A. Because $q_A \in F_A = q_0^B$ and $p_A \in I_A$, and because the transition functions $\cdot_B$ and $\circ_B$ are the extensions of the transition functions $\cdot_A^R$ and, respectively, $\circ_A^R$ to sets, we obtain the following accepting computation in the biautomaton B from the computation shown in Equation (9.1):

$$p_B \in ((((\ldots ((q_0^B \circ_B v_k^R) \cdot_B u_k^R) \ldots) \circ_B v_2^R) \cdot_B u_2^R) \circ_B v_1^R) \cdot_B u_1^R, \qquad (9.2)$$

for a state $p_B \in F_B$. Note that we treat the *deterministic* biautomaton B here like a nondeterministic one. In this way we can apply Lemma 7.17 to the computation from Equation (9.2), to obtain the following computation in B^R:

$$q_0^B \in ((\ldots ((((p_B \cdot_B^R u_1) \circ_B^R v_1) \cdot_B^R u_2) \circ_B^R v_2) \ldots) \cdot_B^R u_k) \circ_B^R v_k. \tag{9.3}$$

Finally, we translate this computation to a computation in the power-set automaton $C = \mathcal{P}(B^R)$ as follows. The initial state of C is $q_0^C = F_B$ and we have $p_B \in F_B$. Further, the transition functions $\cdot_C$ and $\circ_C$ of biautomaton C are extensions of the transition function $\cdot_B^R$ and, respectively, $\circ_B^R$ to sets. This yields the following computation in the deterministic biautomaton C:

$$q_C = ((\ldots ((((q_0^C \cdot_C u_1) \circ_C v_1) \cdot_C u_2) \circ_C v_2) \ldots) \cdot_C u_k) \circ_C v_k.$$

From Equation (9.3) we know that $q_0^B \in q_C$, therefore q_C is an accepting state of C. Altogether we obtain $w \in L(C)$ if and only if $w \in L(A)$. □

9.3 On the Importance of the ⋄- and F-properties

We have just seen that the classical state merging minimization algorithm for deterministic finite automata as well as the elegant Brzozowski minimization algorithm can be generalized to work for deterministic biautomata with the ⋄- and F-property. Now the question arises, how important are these both properties for the minimization algorithms to work properly? First we show that for biautomata in general, where no properties are required, there does not even exist a minimization algorithm. This result is based on Corollary 7.8, which gives a characterization of linear context-free languages in terms of biautomata.

Theorem 9.8. *There is no minimization algorithm converting an arbitrary (deterministic or nondeterministic) biautomaton into an equivalent biautomaton of the same type that has a minimal number of states. This even holds regardless whether the involved biautomata satisfy the F-property or not.*

Proof. Obviously, a minimal deterministic biautomaton with input alphabet Σ that accepts Σ^* has a single state only, which is accepting, and there must be forward and a backward transitions on that single state for every symbol $a \in \Sigma$, regardless whether the F-property is required or not. In case of nondeterministic biautomata the minimal device looks as previously described, or the automaton has forward and backward transitions on that single state, where either the labels on the forward transitions give the set Σ or the labels on the backward transitions give Σ. In this case the F-property needs not to be satisfied. Now assume to the contrary that a minimization algorithm for biautomata exists. Let A be an arbitrary biautomaton. If we apply the minimization algorithm to A, we obtain an equivalent minimal biautomaton B. Then we can check whether B meets one of the above descriptions for a minimal biautomaton accepting Σ^*. If this is so, then $L(A) = \Sigma^*$ and thus A is universal, otherwise $L(A) \neq \Sigma^*$.

Hence we can decide universality for biautomata. This is a contradiction, because for a given linear context-free grammar one can effectively construct an equivalent biautomaton (with the F-property) and *vice versa*, which is shown in Theorems 7.5 and 7.6. But the universality problem for linear context-free grammars is undecidable [46]. This proves that no minimization algorithm for biautomata in general exists. □

In the remainder of this section we stick to deterministic biautomata that satisfy the $\diamond$-property but not the F-property. Our next goal is to show that the F-property is essential for minimizing biautomata by the classical state merging approach or by the Brzozowski-like minimization. We precede our result with the following example.

Example 9.9. Consider the finite language $L = \{aa\}$, which is accepted by the deterministic biautomaton A that is depicted in the left of Figure 9.5. The biautomaton A satisfies both the $\diamond$- and the F-property. In fact, it is easy to see that A is the minimal $\diamond$-F-DBiA for L. Splitting state q_2 and arranging the transitions such that a forward transition can only be followed by a backward transition or *vice versa* results in the biautomaton B, which still satisfies the $\diamond$-property, but now lacks the F-property—see the right of Figure 9.5. It is easy to see that $\mathcal{P}(B^R)$ is isomorphic to B and therefore, $\mathcal{P}([\mathcal{P}(B^R)]^R)$ is isomorphic to B, too. Thus, the Brzozowski-like minimization algorithm does not work properly, because it produces an automaton with five states, but A is equivalent to B and of smaller size.

Next consider the state merging algorithm applied to the deterministic biautomaton B. First, the pairs containing the only accepting state q_3 and some other state are marked as pairs of distinguishable states. Moreover, all pairs of states that contain the sink state are marked as pairs of distinguishable states, too. Then also the state pairs (q_1, q_2) and (q_2', q_3) are marked as pairs of distinguishable states: on taking a backward transition the pair (q_1, q_2) goes to the previously marked pair (q_2', q_3), and with a forward transition the pair (q_1, q_2') goes the marked pair to (q_2, q_3). Finally, the remaining pair (q_2, q_2') is marked as a pair of distinguishable states, because on a forward transition state q_2 goes to the sink state, while state q_2' goes to state q_3. Thus, no states can be merged

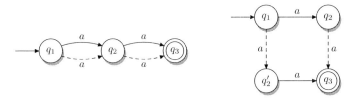

Figure 9.5: Deterministic biautomata A (left) and B (right) accepting the language $L = \{aa\}$. The sink states and corresponding transitions are not shown. Both biautomata satisfy the $\diamond$-property, but only A obeys also the F-property.

in biautomaton B. Therefore, the state merging approach that considers state pairs is not working, if the deterministic biautomaton misses the F-property. $\diamond$

Notice that if the state merging is done by considering the right language of the states, then states q_2 and q_2' from the previous example are identified to be equivalent, because the right language of both states is $\{a\}$. Hence these states can be merged and we obtain a deterministic biautomaton that is of same size as A. Although it looks like that this more general approach on state merging is successful in minimizing deterministic biautomata with the $\diamond$-property, we will show in the following, by a modification of example 9.9, that this is unfortunately not the case.

Theorem 9.10. *There is a deterministic biautomaton with the $\diamond$-property that cannot be minimized by either the state merging algorithm or by the Brzozowski-like minimization algorithm as described earlier.*

Proof. Consider the language $L' = a \cdot (a + b)^* \cdot a$. This language is accepted by the deterministic biautomata C and D depicted in Figure 9.6. Since C is an ordinary deterministic finite automaton, it trivially satisfies the $\diamond$-property, because all backward transitions lead to the non-accepting sink state. But C does not satisfy the F-property. Moreover, also the biautomaton D fulfills the $\diamond$-property, but it does not satisfy the F-property, because for example state q_2 leads with a forward a-transition to a non-accepting state, while with a backward a-transition it goes to the accepting state q_3. It is not hard to see that a state merging algorithm cannot merge any equivalent states in D, because all states are pairwise inequivalent. This can be seen from the right languages of the states, which are as follows:

$$L_D(q_1) = a \cdot (a + b)^* \cdot a, \qquad L_D(q_2) = (a + b)^* \cdot a,$$
$$L_D(q_{2'}) = a \cdot (a + b)^*, \qquad L_D(q_3) = (a + b)^*,$$

and the right language of the sink state (which is not shown in Figure 9.6) is of course the empty set. Since C is of smaller size than D, the state merging approach is not successful for deterministic biautomata with the $\diamond$-property.

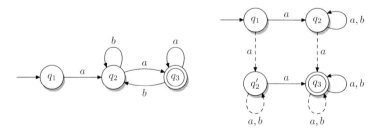

Figure 9.6: Deterministic biautomata C (left) and D (right) accepting the language $L' = a \cdot (a + b)^* \cdot a$. The sink state and corresponding transitions are not shown. Both biautomata satisfy the $\diamond$-property but not the F-property.

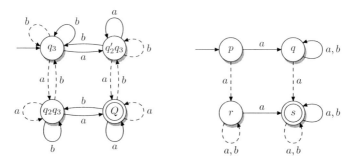

Figure 9.7: Left: deterministic biautomaton $D' = \mathcal{P}(D^R)$ in the intermediate step of the Brzozowski-like minimization approach. The state labels denote the corresponding subset of states, where Q denotes the set $\{q_1, q_2, q_2', q_3\}$. Right: deterministic biautomaton $D'' = \mathcal{P}(D'^R)$ resulting from the final step of the Brzozowski-like minimization approach. The state labels p, q, r, s are abbreviations for the following subsets of the state set $Q = \{q_1, q_2, q_2', q_3\}$ of D': $p = \{Q\}$, $q = \{Q, \{q_2, q_3\}\}$, $r = \{Q, \{q_2', q_3\}\}$, and $s = \{Q, \{q_2, q_3\}, \{q_2', q_3\}, \{q_3\}\}$. As usual, sink states are omitted.

Also the Brzozowski-like minimization method cannot be used to minimize biautomaton D: the left-hand side of Figure 9.7 shows the biautomaton $D' = \mathcal{P}(D^R)$ in the intermediate stage of Brzozowski's algorithm and the right-hand side shows the final biautomaton $D'' = \mathcal{P}([\mathcal{P}(D^R)]^R)$. Notice that D'' is isomorphic to D: states q_1, q_2, q_2', q_3 of biautomaton D can be identified with the states p, q, r, s, respectively, of biautomaton D'.　　□

The straightforward technique to minimize deterministic biautomata that satisfy the $\diamond$-property is to enumerate all possible biautomata of smaller size with the $\diamond$-property and to check language equality. It is not hard to see that this enumeration and the equality check can be done on a deterministic Turing machine in polynomial space. Therefore, one can conclude PSPACE as an upper bound on the complexity of the corresponding minimization problem. Whether this bound is optimal, in terms of complexity theory, is left open. Nevertheless, we conjecture that the minimization problem for deterministic biautomata with the $\diamond$-property is indeed intractable. Evidence that partially supports this conjecture is given by Theorem 8.12, which shows that enforcing the F-property for deterministic biautomata that have the $\diamond$-property may cause an exponential blow-up in the number of states of the biautomaton.

Chapter 10

Lossy Compression of Biautomata

INKING the main topics of Part II and Part III, this chapter revisits the
idea of lossy compression, now applied to biautomata. Chapter 4 in
Part II intensively studied computational complexity aspects of almost-
equivalent and E-equivalent finite automata. These equivalence notions, together with their corresponding concepts of hyper-minimality and, respectively,
E-minimality carry over to biautomata in a natural way. Because Chapter 9 re-
vealed many similarities between classical minimization of finite automata and
of biautomata—at least for such biautomata which satisfy both the $\diamond$-property
and the F-property—one might expect that the computational complexity results from Part II also carry over to the biautomaton case. In fact, we will later
show that the minimization problem for deterministic biautomata that satisfy
the $\diamond$-property and the F-property is NL-complete, just like it is for deterministic finite automata. However, when it comes to lossy compression, the situation
is not that simple.

This chapter is organized as follows. Section 10.1 studies the structure
of almost-equivalent biautomata. On the one hand, we show that some results on almost-equivalent finite automata are also true for almost-equivalent
biautomata. On the other hand, some structural similarities between almost-
equivalent finite automata cannot always be found for biautomata. This lack of
structure allows us to reveal considerable differences between finite automata
and biautomata, and also between hyper-minimization and classical minimization. We show in Section 10.2 that while minimization of deterministic biautomata is NL-complete, the hyper-minimization problem for biautomata becomes NP-complete—recall that both these problems are only NL-complete for
deterministic finite automata. As a consequence, also E-minimization for deterministic biautomata can be shown to be NP-complete.

Throughout this chapter we limit our studies to deterministic biautomata
that satisfy the $\diamond$-property and the F-property. Therefore, whenever in the
following we speak of biautomata, we mean $\diamond$-F-DBiAs.

10.1   On the Structure of Almost-Equivalent Biautomata

Remember that two languages L and L' are *almost-equivalent* ($L \sim L'$) if their symmetric difference $L \triangle L$ is finite. Let us transfer the basic definitions concerning almost-equivalent automata from finite automata to biautomata. Two biautomata A and A' are *almost-equivalent* ($A \sim A'$) if $L(A) \sim L(A')$. A biautomaton A is *hyper-minimal* if there is no biautomaton A', with $A \sim A'$, that has fewer states than A. If q is a state of A and q' a state of A', then these states are *almost-equivalent* ($q \sim q'$) if $L_A(q) \sim L_{A'}(q')$—here the automata A and A' need not be different. Moreover, the notions of preamble and kernel states are defined similar as in the DFA case: let $A = (Q, \Sigma, \cdot, \circ, q_0, F)$ be a deterministic biautomaton. A state of $q \in Q$ is a *kernel state* if it is reachable from the initial state q_0 by an infinite number of inputs ($L_{A,q}$ is infinite) and it is a *preamble state* if it is reachable from q_0 only by a finite number of inputs ($L_{A,q}$ is finite). The *preamble* of A is the set $\mathrm{Pre}(A)$ of all preamble states of A, and the *kernel* of A is the set $\mathrm{Ker}(A)$ of kernel states of A.

Before we study the structure of almost-equivalent biautomata, for comparison reasons we recall the following result of Badr et al. [5] as it was presented in Section 3.1. We will see a similar result for biautomata in Theorem 10.2.

Theorem 3.3. *Let $A = (Q, \Sigma, \delta, q_0, F)$ and $A' = (Q', \Sigma, \delta', q_0', F')$ be two minimal deterministic finite automata with $A \sim A'$. Then there exists a mapping $h \colon Q \to Q'$ satisfying the following conditions.*

1. *If $q \in \mathrm{Pre}(A)$, then $q \sim h(q)$, and if $q \in \mathrm{Ker}(A)$, then $q \equiv h(q)$.*

2. *If $q_0 \in \mathrm{Pre}(A)$, then $h(q_0) = q_0'$, and if $q_0 \in \mathrm{Ker}(A)$, then $h(q_0) \sim q_0'$.*

3. *The restriction of h to $\mathrm{Ker}(A)$ is a bijection between the kernels of A and A', that is compatible with taking transitions:*

   3.a *It is $h(\mathrm{Ker}(A)) = \mathrm{Ker}(A')$, and if $q_1, q_2 \in \mathrm{Ker}(A)$ with $h(q_1) = h(q_2)$, then $q_1 = q_2$.*

   3.b *It is $h(\delta(q, a)) = \delta'(h(q), a)$, for all $q \in \mathrm{Ker}(A)$ and all $a \in \Sigma$.*

Moreover, if both automata A and A' are hyper-minimal, then also the following condition holds.

4. *The restriction of h to $\mathrm{Pre}(A)$ is a bijection between the preambles of A and A', that is compatible with taking transitions, except for transitions from preamble to kernel states:*

   4.a *It is $h(\mathrm{Pre}(A)) = \mathrm{Pre}(A')$, and if $q_1, q_2 \in \mathrm{Pre}(A)$ with $h(q_1) = h(q_2)$, then $q_1 = q_2$.*

   4.b *It is $h(\delta(q, a)) = \delta'(h(q), a)$, for all $q \in \mathrm{Pre}(A)$ and all $a \in \Sigma$, that satisfy $\delta(q, a) \in \mathrm{Pre}(A)$.*

Now let us investigate, which of the structural similarity results for almost-equivalent hyper-minimal DFAs carry over to biautomata. We first show that a result similar to Lemma 3.4 from Part II also holds for deterministic biautomata.

Lemma 10.1. *Let* $A = (Q, \Sigma, \cdot, \circ, q_0, F)$ *and* $A' = (Q', \Sigma, \cdot', \circ', q_0', F')$ *be two deterministic biautomata, both satisfying the $\diamond$-property and the F-property. Let* $q \in Q$ *and* $q' \in Q'$*, then* $q \sim q'$ *if and only if* $(q \cdot u) \circ v \sim (q' \cdot' u) \circ' v$*, for all words* $u, v \in \Sigma^*$*. Moreover,* $q \sim q'$ *even implies* $(q \cdot u) \circ v \equiv (q' \cdot' u) \circ' v$*, for all words* $u, v \in \Sigma^*$ *with* $|uv| \geq k = |Q \times Q'|$*.*

Proof. Assume there are words u and v such that the states $p = (q \cdot u) \circ v$ and $p' = (q' \cdot' u) \circ' v$ are not almost-equivalent. This means that there are infinitely many words $w \in L_A(p) \triangle L_{A'}(p')$, which implies that there are infinitely many words $uwv \in L_A(q) \triangle L_{A'}(q')$. This shows that states q and q' can only be almost-equivalent if $(q \cdot u) \circ v \sim (q' \cdot' u) \circ' v$ holds for all $u, v \in \Sigma^*$. The reverse implication is trivial: if $(q \cdot u) \circ v \sim (q' \cdot' u) \circ' v$, for all $u, v \in \Sigma^*$, then we obtain $q \sim q'$ by choosing $u = v = \lambda$. This proves the first part of the Lemma.

For the second part let $u = a_1 a_2 \ldots a_\ell$ and $v = a_m a_{m-1} \ldots a_{\ell+1}$, with $a_i \in \Sigma$ for $1 \leq i \leq m$ and $m \geq k$. Assume $q \sim q'$ and consider the state pairs (q_i, q_i'), for $0 \leq i \leq m$, the automata visit in their computations $(q \cdot u) \circ v$ and $(q' \cdot' u) \circ' v$:

$$(q_i, q_i') = \begin{cases} (q, q'), & \text{for } i = 0, \\ (q_{i-1} \cdot a_i, \, q_{i-1}' \cdot' a_i) & \text{for } 1 \leq i \leq \ell, \\ (q_{i-1} \circ a_i, \, q_{i-1}' \circ' a_i) & \text{for } \ell+1 \leq i \leq m. \end{cases}$$

Since $m \geq |Q \times Q'|$, there must be integers i and j, with $0 \leq i < j \leq m$, for which we have $(q_i, q_i') = (q_j, q_j')$. If $j \leq \ell$, then u can be written as $u = u_1 u_2 u_3$ such that

$$q \cdot u_1 = q_i, \qquad q_i \cdot u_2 = q_i, \qquad (q_i \cdot u_3) \circ v = p,$$
$$q' \cdot' u_1 = q_i', \qquad q_i' \cdot' u_2 = q_i', \qquad (q_i' \cdot' u_3) \circ' v = p'.$$

If the states $p = (q \cdot u) \circ v$ and $p' = (q' \cdot' u) \circ' v$ are not equivalent, then there is a word $w \in L_A(p) \triangle L_{A'}(p')$. It follows that $u_1 u_2^n u_3 wv \in L_A(q) \triangle L_{A'}(q')$, for all $n \geq 0$. This is a contradiction to $q \sim q'$. If $\ell + 1 \leq i$, then we can find a similar partition of the word $v = v_3 v_2 v_1$, such that a word w in the symmetric difference of the two states $p = (q \cdot u) \circ v$ and $p' = (q' \cdot' u) \circ' v$ induces infinitely many words $uwv_3 v_2^n v_1 \in L_A(q) \triangle L_{A'}(q')$, for all $n \geq 0$. It remains to discuss the case $i \leq \ell < j$. Now the words u and v can be written as $u = u_1 u_2$ and $v = v_2 v_1$ such that

$$q \cdot u_1 = q_i, \qquad (q_i \cdot u_2) \circ v_1 = q_i, \qquad q_i \circ v_2 = p,$$
$$q' \cdot' u_1 = q_i', \qquad (q_i' \cdot' u_2) \circ' v_1 = q_i', \qquad q_i' \circ' v_2 = p'.$$

Now, if the states $p = (q \cdot u) \circ v$ and $p' = (q' \cdot' u) \circ' v$ are not equivalent, then there is a word $w \in L_A(p) \triangle L_{A'}(p')$, and it follows that $u_1 u_2^n w v_2 v_1^n \in L_A(q) \triangle L_{A'}(q')$, for all $n \geq 0$. This is again a contradiction to $q \sim q'$, hence the two states $(q \cdot u) \circ v$ and $(q' \cdot' u) \circ' v$ must be equivalent. $\qquad \square$

Now we come to a mapping between the states of two almost-equivalent biautomata. As in the case of finite automata, we can find an isomorphism between the kernels of the two automata. However, we cannot find a similar isomorphism between their preambles. Of course, two almost-equivalent hyper-minimal $\diamond$-F-DBiAs must have the same number of states, and if their kernels are isomorphic, then also their preambles must be of same size. But still we cannot always find a bijective mapping that preserves almost-equivalence, as in the case of finite automata. We will later see an example for this phenomenon, but first we present our result on the structural similarity between almost-equivalent minimal $\diamond$-F-DBiAs, which should be compared to Theorem 3.3.

Theorem 10.2. *Let $A = (Q, \Sigma, \cdot, \circ, q_0, F)$ and $A' = (Q', \Sigma, \cdot', \circ', q_0', F')$ be two minimal deterministic biautomata satisfying the $\diamond$-property and the F-property, such that $A \sim A'$. Then there exists a mapping $h \colon Q \to Q'$ that satisfies the following conditions.*

1. *If $q \in \mathrm{Pre}(A)$, then $q \sim h(q)$, and if $q \in \mathrm{Ker}(A)$, then $q \equiv h(q)$.*

2. *If $q_0 \in \mathrm{Pre}(A)$, then $h(q_0) = q_0'$, and if $q_0 \in \mathrm{Ker}(A)$, then $h(q_0) \sim q_0'$.*

3. *The restriction of h to $\mathrm{Ker}(A)$ is a bijection between the kernels of A and A', that is compatible with taking transitions:*

   3.a *It is $h(\mathrm{Ker}(A)) = \mathrm{Ker}(A')$, and if $q_1, q_2 \in \mathrm{Ker}(A)$ with $h(q_1) = h(q_2)$, then $q_1 = q_2$.*

   3.b *It is $h(q \cdot a) = h(q) \cdot' a$ and $h(q \circ a) = h(q) \circ' a$, for all $q \in \mathrm{Ker}(A)$ and all $a \in \Sigma$.*

Proof. In order to define the mapping h we choose for every state $q \in Q$ two words u_q and v_q as follows. If q is a kernel state of A, then there are infinitely many words $u, v \in \Sigma^*$ such that $(q_0 \cdot u) \circ v = q$. Hence we can choose for every kernel state $q \in \mathrm{Ker}(A)$ two words u_q and v_q with $|u_q v_q| \geq k = |Q \times Q'|$. More-over, for every preamble state $q \in \mathrm{Pre}(A)$, we also fix some shortest words u_q and v_q (that is, where $|u_q v_q|$ is shortest possible) such that $(q_0 \cdot u_q) \circ v_q = q$. The mapping $h \colon Q \to Q'$ is then defined by $h(q) = (q_0' \cdot' u_q) \circ' v_q$. In the following we show that this mapping satisfies the statements of the theorem. Since $A \sim A'$ it must be $q_0 \sim q_0'$. If $q_0 \in \mathrm{Pre}(A)$, then we have $u_{q_0} = v_{q_0} = \lambda$, and so $h(q_0) = q_0'$, which proves one part of statement 2. From Lemma 10.1 we obtain $(q_0 \cdot u_q) \circ v_q \sim (q_0' \cdot' u_q) \circ' v_q$, for all preamble states $q \in \mathrm{Pre}(A)$, and further $(q_0 \cdot u_q) \circ v_q \equiv (q_0' \cdot' u_q) \circ' v_q$, for all kernel states $q \in \mathrm{Ker}(A)$. This proves statement 1. Now, if $q_0 \in \mathrm{Ker}(A)$, then $q_0' \sim q_0 \equiv h(q_0)$, which proves the other part of statement 2. Moreover, for all kernel states $q \in \mathrm{Ker}(A)$, we have $q \equiv h(q)$, which implies

$$h(q \cdot a) \equiv q \cdot a \equiv h(q) \cdot' a \qquad \text{and} \qquad h(q \circ a) \equiv q \circ a \equiv h(q) \circ' a,$$

for all symbols $a \in \Sigma$. Since A' is a minimal biautomaton, it does not contain a pair of different but equivalent states. Therefore it must be $h(q \cdot a) = h(q) \cdot' a$ and $h(q \circ a) = h(q) \circ' a$, for all $a \in \Sigma$, which proves statement 3.b.

It remains to prove statement 3.a, namely that h is bijective between the kernels of the two biautomata. If $q \in \mathrm{Ker}(A)$, then $|u_q v_q| \geq |Q \times Q'| \geq |Q'|$, hence state $h(q)$ must be a kernel state of A'. So $h(\mathrm{Ker}(A)) \subseteq \mathrm{Ker}(A')$. Moreover, the mapping h is injective: if $h(p) = h(q)$, then we know $p \equiv h(p) = h(q) \equiv q$, but since A is a minimal biautomaton, this implies $p = q$. Therefore we have $|\mathrm{Ker}(A)| \leq |\mathrm{Ker}(A')|$. By exchanging the roles of A and A' we can also find an injective mapping $h' \colon Q' \to Q$ with $h(\mathrm{Ker}(A')) \subseteq \mathrm{Ker}(A)$, which shows that $|\mathrm{Ker}(A')| \leq |\mathrm{Ker}(A)|$. Altogether we obtain $|\mathrm{Ker}(A)| = |\mathrm{Ker}(A')|$, so the mapping h must also be surjective on $\mathrm{Ker}(A')$. This concludes our proof. □

Notice that Theorem 10.2 requires the two almost-equivalent biautomata A and A' to be minimal but not necessarily hyper-minimal. Of course, the theorem also holds for hyper-minimal automata, since these are always minimal. However, the question is whether we can find more structural similarities—like statement 4 from Theorem 3.3 on DFAs—if both biautomata are hyper-minimal. Unfortunately the answer is no, as the following example demonstrates.

Example 10.3. Consider the biautomaton A which is depicted on top in Figure 10.1 on page 161, where as usual, transitions which are not shown lead to a non-accepting sink state, which is also not shown. The state labels of the eight states in the lower two rows of the automaton denote the right languages of the respective states. The kernel of A consists of those states and the sink state. The right languages of the states q_0, q_1, and q_2, which constitute the preamble of A, are as follows:

$$L_A(q_0) = L(A) = (a + b)ba^* b + c^* a, \quad L_A(q_1) = ba^* b + \lambda, \quad L_A(q_2) = ba^* b.$$

One can verify that A satisfies the $\diamond$- and the F-property. Let us first show that A is hyper-minimal.

Claim. The biautomaton A from Figure 10.1 is hyper-minimal.

Proof. First notice that the right languages of all states of A are pairwise distinct, so all states of A are pairwise non-equivalent. Hence, by Theorem 9.3 the biautomaton A is minimal. Assume B is a hyper-minimal biautomaton that is almost-equivalent to A. We have to show that B has at least as many states as A. We know from Theorem 10.2 that the kernels of A and B are isomorphic, hence it suffices to show that B has at least three states in its preamble. Let us denote the initial state of B by q_0^B and its forward and backward transition functions by $\cdot_B$ and $\circ_B$, respectively. We have $q_0^B \sim q_0$ because $B \sim A$. By Lemma 10.1 we obtain $(q_0^B \cdot_B u) \circ_B v \sim (q_0 \cdot u) \circ v$, for all words u and v. Since state q_0 is not almost-equivalent to any kernel state of A, also state q_0^B is not a kernel state of B either—remember that the kernels of A and B are isomorphic. Further, also the state $q_1 = q_0 \cdot a$ is not almost-equivalent to any kernel state, and not almost-equivalent to q_0, therefore the state $q_0^B \cdot_B a$ must be a another preamble state in B, too. Let us denote this state by q_1^B.

If we can show that B has another preamble state, then we know that A is hyper-minimal. Therefor assume for the sake of contradiction that q_0^B and q_1^B are the only preamble states of B. Then state q_1^B is the only state of B that is almost-equivalent to state q_1 of automaton A. Since $q_1 \sim' q_2$, state q_1^B is also the only state of B that is almost-equivalent to q_2. By Lemma 10.1, state $q_2 = q_0 \cdot b$ of A must be almost-equivalent to state $q_0^B \cdot_B b$ of B, so we conclude that $q_0^B \cdot_B b = q_1^B = q_0^B \cdot_B a$ holds. Now we consider two cases, namely whether q_1^B is an accepting state or not.

If $q_1^B = q_0^B \cdot_B a$ is not accepting, then also the state $q_0^B \circ_B a$ must not be accepting, due to the F-property of B. However, state $q_0^B \circ_B a$ must be almost-equivalent to state $q_0 \circ a$, which is the *accepting* kernel state c^* in A. Also the corresponding kernel state of B must be accepting, therefore this cannot be the target of the transition $q_0^B \circ_B a$. Since there is no other kernel state which is almost-equivalent to c^*, we conclude that the automaton B must have yet another preamble state $q_0^B \circ_B a$, different from q_0^B and q_1^B—a contradiction.

The other case is similar: if $q_1^B = q_0^B \cdot_B b$ is an accepting state, then also state $q_0^B \circ_B b$ must be accepting. Moreover, this state must be almost-equivalent to the kernel state $q_0 \circ b$, that is, the state $(a+b)ba^*$ of A. The corresponding kernel state of B is also not accepting, so it cannot be the target of the transition $q_0^B \circ_B b$. Again, there is no other kernel state that is almost-equivalent to state $(a+b)ba^*$, so B must possess another preamble state, different from q_0^B and q_1^B. This concludes the proof of the claim. $\square$

Now consider the biautomaton A', depicted on the bottom of Figure 10.1. This biautomaton accepts the language $L(A') = (a + b)ba^*b + cc^*a$, so it is almost-equivalent to A. Since A and A' have the same number of states, and A is hyper-minimal, the automaton A' is hyper-minimal, too. Consider a mapping h from the states of A to the states of A', that satisfies the conditions of Theorem 10.2. Between the kernels of the automata, the mapping is clear. Moreover, since q_0 and q_0' are preamble states, it must be $h(q_0) = q_0'$. This can even be concluded if q_0 and q_0' were not the initial states, because state q_0' of A' is the only state that is almost-equivalent to q_0, and h must satisfy $q \sim h(q)$ for all states q. With the same argumentation we obtain $h(q_1) = h(q_2) = q_2'$: state q_2' is the only state of A' that is almost-equivalent to states q_1 and q_2 of A. The mapping h is now fully defined, so in this example, there is no other possible mapping from the states of A to the states of A' that preserves almost-equivalence.

Notice that the mapping h is not a bijection between the preambles: because we have $h(q_1) = h(q_2) = q_2'$, it is not injective, and it neither is surjective, since no state of A is mapped to state q_1' of B. This shows that the bijection condition 4.a of Theorem 3.3 for preambles of deterministic finite automata does not hold for biautomata.

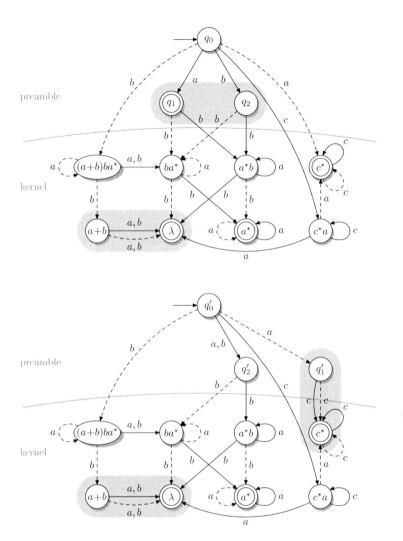

Figure 10.1: Two almost-equivalent hyper-minimal deterministic biautomata with the $\diamond$-property and the F-property. Biautomaton A (on the top) accepts the language $L(A) = (a + b)ba^*b + c^*a$ and A' (on the bottom) accepts the language $L(A') = (a + b)ba^*b + cc^*a$. The preambles of these biautomata are $\text{Pre}(A) = \{q_0, q_1, q_2\}$ and $\text{Pre}(A') = \{q'_0, q'_1, q'_2\}$. The states q_1, q_2, q'_1, and q'_2 have the following right languages: $L_A(q_1) = ba^*b + \lambda$, $L_A(q_2) = L_{A'}(q'_2) = ba^*b$, and $L_{A'}(q'_1) = cc^*$. The gray shading of state pairs denotes almost-equivalence, that is, we have $q_1 \sim q_2$, $a + b \sim \lambda$, and $q'_1 \sim c^*$.

Similarly, also the relation between transitions in the preambles of the automata as described in Condition 4.b of Theorem 3.3 cannot be found here. To see this, we have to look at the converse mapping from the states of A' to states of A. Denote this mapping by $h' \colon Q' \to Q$ and consider the states q'_0 and q'_2 in the preamble of B. The image of q'_0 under h' must be $h'(q'_0) = q_0$. For the image of q'_2 we have two possibilities, namely state q_1 or state q_2. Assume we have $h'(q'_2) = q_2$. Notice that we have the transition $q'_0 \cdot' a = q'_2$ between two preamble states of A'. However, this transition is not preserved by the mapping h': we have $h'(q'_0 \cdot' a) = h'(q'_2) = q_2$ but $h'(q'_0) \cdot a = q_0 \cdot a = q_1$. If instead of q_2 we choose q_1 as the image of q'_2, then we obtain a similar mismatch by $h'(q'_0 \cdot' b) = h'(q'_2) = q_1$ but $h'(q'_0) \cdot b = q_0 \cdot b = q_2$.

Of course there exist bijections between the state sets of the biautomata A and A'. But none of these can preserve almost-equivalence, because the corresponding almost-equivalence classes in the state sets are not always of same size. For example, there are two states in biautomaton A that are almost-equivalent to q_1, namely q_1 itself and q_2, but in biautomaton A' there is only state q'_2 in its equivalence class. $\diamond$

10.2 Computational Complexity of Biautomata Minimization Problems

Given a deterministic finite automaton, it is an easy task to construct an equivalent minimal automaton. A lot of minimization algorithms for DFAs are known, the most efficient of them being Hopcroft's algorithm [61] with a running time of $O(n \log n)$, where n is the number of states of the input DFA. In fact, the decision version of the DFA minimization problem—given a DFA A and an integer n, decide whether there exists an n-state DFA B with $A \equiv B$—is NL-complete [22].

Concerning minimization of deterministic biautomata, it was discussed in Chapter 9 how classical DFA minimization techniques can also be applied to biautomata. In the following we investigate the computational complexity of the minimization problem for deterministic biautomata and show that it is NL-complete, too. For proving NL-hardness we give a reduction from the following NL-complete variant[1] of the graph reachability problem [69, 70]:

> given a directed graph $G = (V, E)$ with vertices $V = \{v_1, v_2, \dots, v_n\}$, where every vertex has at most two successors and at most two predecessors, decide whether v_n is reachable from v_1.

Now we can show the following result.

---

[1] The general graph reachability problem can be reduced to the case where every vertex has at most two successors by appending after each vertex that has more than two successors a small tree-like subgraph to simulate the multiple outgoing edges. A similar construction can be used to reduce the number of predecessors, to obtain a graph where also the number of predecessors of a vertex is at most two.

Theorem 10.4 (Minimization Problem). *The problem of deciding for a given deterministic biautomaton A, which satisfies the $\diamond$-property and the F-property, and an integer n, whether there exists an equivalent deterministic biautomaton B with n states, satisfying the $\diamond$- and F-property, is NL-complete.*

Proof. For the NL upper bound we use the following algorithm for computing the number k of equivalence classes of the state set of A. Let $q_1, q_2, \ldots q_m$ be some fixed order of the states of A and initially set $k = 0$. For all states q_i (in ascending order) do the following: if $q_i \not\equiv q_j$, for all $j < i$, then increment k. Finally, if $k \leq n$, then the answer is yes, otherwise it is no. Because A has the $\diamond$-property and the F-property, due to Corollary 6.11 it suffices to consider only forward transitions to decide whether $q_i \not\equiv q_j$. Therefore, in order to check whether $q_i \not\equiv q_j$ holds, we can check equivalence of the two DFAs obtained from A by making q_i and q_j, respectively, the initial state and considering only forward transitions. Since equivalence of DFAs can be decide in NL, the whole algorithm can be implemented on a nondeterministic logspace Turing machine.

For NL-hardness we give a reduction from the above described variant of the graph reachability problem. The idea is to transform the graph into a DFA A_1 that accepts the empty language if the instance of the graph reachability problem is a "no" instance, and it accepts a non-empty, non-universal language otherwise. Then an equivalent biautomaton A is built by a cross-product construction of A_1 and the reverse of A_1. The biautomaton A is equivalent to a single-state biautomaton if and only if the graph reachability problem is a "no" instance. The challenge in this approach is to make sure that the construction of the reverse of A_1 can be carried out by a logarithmic space-bounded Turing machine, because in general this construction induces an exponential blow-up in the number of states of the automaton. Therefore we construct the DFA A_1 such that its reversal is also a deterministic automaton.

Let $G = (V, E)$, with $V = \{v_1, v_2, \ldots, v_n\}$ be a directed graph where every vertex has at most two successors and at most two predecessors. We construct the partial DFA $A_1 = (Q, \Sigma, \delta, q_0, F)$—partial means that some transitions may be undefined—over the alphabet $\Sigma = \{a, b\}$: the state set consists of the vertices and edges of G, that is, $Q = V \cup E$, the initial state is $q_0 = v_1$, and set of final states is $F = \{v_n\}$. The transitions in states $v_i \in V$ are defined as follows:

- if v_i has two successors v_{j_1} and v_{j_2}, that is, if $(v_i, v_{j_1}), (v_i, v_{j_2}) \in E$ and $j_1 < j_2$, then $\delta(v_i, a) = (v_i, v_{j_1})$ and $\delta(v_i, b) = (v_i, v_{j_2})$,

- if v_i has one successor v_j, that is, if (v_i, v_j) is the only edge in E with v_i on the left-hand side, then $\delta(v_i, a) = (v_i, v_j)$,

Finally, the transitions in states $(v_i, v_j) \in E$ are defined as follows:

- if (v_i, v_j) is the only edge in E with vertex v_j on the right-hand side, then $\delta((v_i, v_j), a) = v_j$,

- if there are two edges (v_{i_1}, v_j) and (v_{i_2}, v_j) in E with vertex v_j on the right-hand side, then $\delta((v_{i_1}, v_j), a) = v_j$ and $\delta((v_{i_2}, v_j), b) = v_j$.

All other transitions are undefined. Note that every state $(v_i, v_j) \in E$ has exactly one outgoing and one ingoing transition and every state $v_i \in V$ has at most one outgoing and at most one ingoing transition for each alphabet symbol. Therefore, the reverse automaton $A_1^R = (Q, \Sigma, \delta^R, v_n, \{v_1\})$, where all transitions are reversed and the initial and (single) final state are interchanged, is also a partial DFA. The biautomaton A can now be constructed by a cross-product construction, simulating A_1 in the first component using its forward transitions, and simulating A_1^R in the second component using backward transitions—see [74] for details of this construction. Clearly this construction can be realized by a logarithmic space-bounded deterministic Turing machine.

It remains to prove the correctness of the reduction. First assume that in the graph G the vertex v_n is not reachable from v_1. Then clearly language $L(A_1) = L(A)$ is empty, so there exists a single-state biautomaton B that is equivalent to A. Next assume v_n is reachable from v_1 in G. Then clearly the language $L(A) = L(A_1)$ is not empty, and because $v_1 \neq v_n$ it is also not equal to Σ^*. Therefore there exists no single-state biautomaton B that is equivalent to A. Since we know $\mathsf{NL} = \mathsf{coNL}$ from the results of Immerman [65] and Szelepcsényi [113], the theorem is proven. □

Now we turn to hyper-minimization. For deterministic finite automata the situation is similar as in the case of classical minimization: efficient hyper-minimization algorithms with running time $O(n \log n)$ are given by Gawrychowski and Jeż [34] and by Holzer and Maletti [59], and it was shown in Section 4.2 that the hyper-minimization problem for DFAs is NL-complete. On the one hand, since classical DFA minimization methods also work well for biautomata, one could expect that hyper-minimization of deterministic biautomata is as easy as for DFAs. On the other hand, the problems related to the structure of hyper-minimal biautomata, which we discussed in Section 10.1, already give hints that hyper-minimization of biautomaton may not be so easy. In fact, we show in the following that the hyper-minimization problem for deterministic biautomata is NP-complete. To prove NP-hardness we give a reduction from the NP-complete $\mathsf{MAX}\text{-}2\text{-}\mathsf{SAT}$ problem which is defined as follows—see the book of Papadimitriou [102]:

> given a Boolean formula φ in conjunctive normal form, where each clause has exactly two literals, and an integer k, decide whether there is a truth assignment that satisfies at least k clauses of φ.

Before we give a detailed proof of NP-hardness, we want to describe the key idea of the reduction. Given as instance of $\mathsf{MAX}\text{-}2\text{-}\mathsf{SAT}$ a formula φ and number k, we construct a biautomaton A_φ, such that for every clause that can be satisfied in φ we can save one state of A_φ, obtaining an almost-equivalent biautomaton. Every clause of φ will be translated to a part of the biautomaton using a separate alphabet, so that the clause gadgets in A_φ are mostly independent from each other. Assume that $\varphi_i = (\ell_{i_1} \vee \ell_{i_2})$ is a clause of φ, its first literal is $\ell_{i_1} = x_u$, and

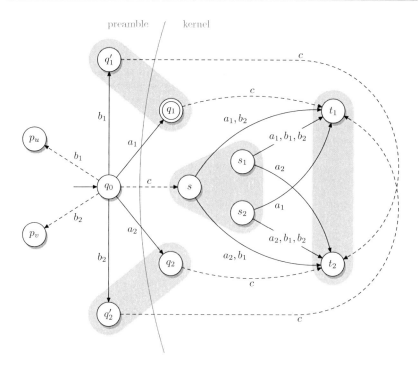

preamble / kernel

Figure 10.2: Simplified structure of the part of biautomaton A_φ corresponding to the clause $\varphi_i = (x_u \vee \overline{x}_v)$. The gray shading denotes almost-equivalence of states.

the second is $\ell_{i_2} = \overline{x}_v$, for some variables x_u and x_v. Then the biautomaton A_φ contains the structure which is depicted in Figure 10.2.

The states q_1 and q_1' correspond to literal ℓ_{i_1}, and states q_2 and q_2' to the literal ℓ_{i_2}. States p_u and p_v correspond to the variables x_u and x_v, respectively, and are shared by all clause gadgets related to these variables. Now assume that there is a truth assignment χ to the variables that satisfies clause φ_i, say by $\chi(x_u) = 1$. Then we make the preamble state p_u accepting, and merge the preamble state q_1' to the almost-equivalent state q_1. To preserve the $\diamond$-property of the automaton, we further re-route the backward c transition of the initial state, making state s_1 the target of the transition. The case where the clause φ_i is satisfied by the second literal corresponds to the similar situation, where state p_v stays non-accepting, state q_2' is merged to q_2, and the target of the backward c transition from q_0 is state s_2. The changing of acceptance of preamble states only introduces a finite number of errors. Further, the merging of preamble states to almost-equivalent kernel states also yields an almost-equivalent automaton.[2] Therefore, if k clauses of φ can be satisfied, then k states of A_φ can be saved.

The other direction, that is, the deduction of a truth assignment χ from a given automaton B that is almost-equivalent to A_φ, is similar: let $\chi(x_u) = 1$ if and only if state p_u of automaton B is accepting. Now assume that state B has k states less than A_φ. The reduction will make sure that only the states q'_1 and q'_2 can be saved. If for example state q'_1 is not present in B, then the initial state of B must enter state q_1 on reading symbol b_1 with a forward transition. Due to the F-property of B, the state p_u reached from the initial state by taking a backward b_1 transition must be accepting. Since the variable states are shared by all clause gadgets, the information that p_u is accepting—which means that variable x_u should be assigned truth value 1—is transported to all other clause gadgets that use variable x_u. Therefore, no state corresponding to the negative literal $\overline{x}_u$ can be saved, that is, no clause can be satisfied by a literal $\overline{x}_u$.[3] It may be the case that both states q'_1 and q'_2 are merged to their almost-equivalent kernel states q_1 and q_2, respectively. But then, due to the $\diamond$-property, the initial state must go to some state s' on reading a c symbol with a backward transition. This state s' must go to state t_1 on a forward b_1 transition and it must go to state t_2 on a forward b_2 transition. Such a state is not present in the automaton A_φ, so this state s' is an additional state in the preamble of B. Hence, even if both states q'_1 and q'_2 are merged into the kernel, the clause gadget in B cannot save more than one state compared to the clause gadget in A_φ. Altogether, for every state that B has less than A, there is a clause of φ that is satisfied by χ.

We now present our result on the **NP**-hardness of the hyper-minimization problem with a detailed proof.

Theorem 10.5. *The problem of deciding for a given deterministic biautomaton A, that satisfies the $\diamond$-property and the F-property, and an integer n, whether there exists an almost-equivalent deterministic biautomaton B with n states, satisfying the $\diamond$- and F-property, is* **NP***-hard.*

Proof. Let φ and k form an instance of the **MAX-2-SAT** problem, where k is an integer and $\varphi = \varphi_1 \wedge \varphi_2 \wedge \cdots \wedge \varphi_m$ is a Boolean formula in conjunctive normal-form over the set of variables $X = \{x_1, x_2, \ldots, x_n\}$, where each clause φ_i

---

[2] In general, this needs some more argumentation. Here the described changes in the automaton preserved the $\diamond$-property, and the F-property. Therefore the languages accepted by the original and the modified biautomaton are the same as the languages accepted by the contained DFAs (using only forward transitions). Now almost-equivalence of these DFAs, and thus, of the biautomata, follows from the fact that merging preamble states to almost-equivalent kernel states in a DFA preserves almost-equivalence.

[3] The reader may have noticed that there is still a possibility to "cheat": one could use accepting and non-accepting copies of variable states in the preamble in order to satisfy a lot more clauses than possible. We take care of this problem in the detailed proof. (The problem can be solved by using many copies $b_{1,j}$ and $b_{2,j}$ of the b_1 and b_2 symbols, each connected to a different copy $p_{u,j}$ of variable states. If the number of these copies is larger than the number of clauses, then the "cheat" turns out to be a bad trade-off.)

contains exactly two literals ℓ_{i_1} and ℓ_{i_2}. Instead of directly constructing the bi-automaton A_φ for the instance of the hyper-minimization problem, we describe the language L_φ accepted by this biautomaton. The integer k' for the instance of the hyper-minimization problem will be the number of states of A_φ minus k, so for each clause that has to be satisfied in φ a state of A_φ is to be saved. From the definition of L_φ one can see that A_φ can indeed be constructed from φ in polynomial time. After the definition of L_φ, we will analyze the structure of A_φ and prove the correctness of the described reduction.

Let us define the language L_φ. The alphabet Σ, over which L_φ is defined, is

$$\Sigma = \{\$\} \cup \bigcup_{i=1}^{m} \Sigma^{(i)} \cup \bigcup_{j=1}^{n} \bigcup_{h=1}^{m+1} \{\#_{j,h}\},$$

where for $1 \leq i \leq m$ the alphabet $\Sigma^{(i)}$ is

$$\Sigma^{(i)} = \{a_1^{(i)}, a_2^{(i)}\} \cup B_1^{(i)} \cup B_2^{(i)} \cup \{c^{(i)}, d^{(i)}, e_1^{(i)}, e_2^{(i)}, f_1^{(i)}, f_2^{(i)}, f_3^{(i)}, f_4^{(i)}, f_5^{(i)}\},$$

with

$$B_1^{(i)} = \bigcup_{h=1}^{m+1} \{b_{1,h}^{(i)}\} \qquad \text{and} \qquad B_2^{(i)} = \bigcup_{h=1}^{m+1} \{b_{2,h}^{(i)}\}.$$

The language L_φ consists of *clause languages* L_{φ_i} and *variable languages* L_{x_j}:

$$L_\varphi = \bigcup_{i=1}^{m} L_{\varphi_i} \cup \bigcup_{j=1}^{n} L_{x_j}.$$

The variable languages L_{x_j}, for $1 \leq j \leq n$, are defined as follows:

$$L_{x_j} = \bigcup_{h=1}^{m+1} \left(\{\#_{j,h}\} \cdot \{\$\}^* \cdot \{\#_{j,h}\} \cdot B_{x_j,h} \right),$$

where the set $B_{x_j,h}$ contains the symbol $b_{1,h}^{(i)}$ ($b_{2,h}^{(i)}$) if and only if the first (second, respectively) literal in clause φ_i corresponds to the variable x_j. More formally,

$$B_{x_j,h} = \{ b_{1,h}^{(i)} \mid \varphi_i = (\ell_{i_1} \vee \ell_{i_2}) \text{ and } \ell_{i_1} \in \{x_j, \overline{x}_j\} \}$$

$$\cup \{ b_{2,h}^{(i)} \mid \varphi_i = (\ell_{i_1} \vee \ell_{i_2}) \text{ and } \ell_{i_2} \in \{x_j, \overline{x}_j\} \}.$$

For $1 \leq i \leq m$, the clause language L_{φ_i} is defined over the alphabet $(\{\$\} \cup \Sigma^{(i)})^*$ as follows—for a better readability we omit the upper index (i) for symbols from $\Sigma^{(i)}$ in the following descriptions:

- If $\varphi_i = (x_{i_1} \vee x_{i_2})$, then the language L_{φ_i} is defined by the following regular expression:

$$L_{\varphi_i} = a_1 \cdot (e_1^* + d^+c) + B_1 \cdot (e_1^+ + d^*c) + a_2 \cdot (e_2^* + d^+c) + B_2 \cdot (e_2^+ + d^+c)$$
$$+ \$^+ \cdot \Big(f_1 \cdot (e_1^* + d^+c) \cdot f_1 + f_2 \cdot (e_2^* + d^*c) \cdot f_2$$
$$+ f_3 \cdot ((a_1 + B_2) \cdot d^+ + (a_2 + B_1) \cdot d^*) \cdot f_3$$
$$+ f_4 \cdot ((a_1 + B_1 + B_2) \cdot d^+ + a_2 \cdot d^*) \cdot f_4$$
$$+ f_5 \cdot (a_1 \cdot d^+ + (a_2 + B_1 + B_2) \cdot d^*) \cdot f_5 \Big) \cdot \$^+.$$

- If $\varphi_i = (x_{i_1} \vee \overline{x}_{i_2})$, then L_{φ_i} is defined similar as in the first case, but we use $(e_2^+ + d^*c)$ instead of $(e_2^* + d^*c)$: (parts which are the same as in the first case are shown grayed out)

$$L_{\varphi_i} = a_1 \cdot (e_1^* + d^+c) + B_1 \cdot (e_1^+ + d^*c) + a_2 \cdot (e_2^+ + d^*c) + B_2 \cdot (e_2^+ + d^+c)$$
$$+ \$^+ \cdot \Big(f_1 \cdot (e_1^* + d^+c) \cdot f_1 + f_2 \cdot (e_2^+ + d^*c) \cdot f_2$$
$$+ f_3 \cdot ((a_1 + B_2) \cdot d^+ + (a_2 + B_1) \cdot d^*) \cdot f_3$$
$$+ f_4 \cdot ((a_1 + B_1 + B_2) \cdot d^+ + a_2 \cdot d^*) \cdot f_4$$
$$+ f_5 \cdot (a_1 \cdot d^+ + (a_2 + B_1 + B_2) \cdot d^*) \cdot f_5 \Big) \cdot \$^+.$$

- If $\varphi_i = (\overline{x}_{i_1} \vee x_{i_2})$, then L_{φ_i} is defined as in the first case, but we use $(e_1^+ + d^+c)$ instead of $(e_1^* + d^+c)$.

- If $\varphi_i = (\overline{x}_{i_1} \vee \overline{x}_{i_2})$, then L_{φ_i} is defined as in the first case, but we use $(e_1^+ + d^+c)$ instead of $(e_1^* + d^+c)$, and $(e_2^+ + d^*c)$ instead of $(e_2^* + d^*c)$.

This concludes the definition of the language L_φ.

Now let $A_\varphi = (Q, \Sigma, \cdot, \circ, q_0, F)$ be the canonical biautomaton for L_φ, with the state set $Q = \{\, u^{-1}L_\varphi v^{-1} \mid u, v \in \Sigma^* \,\}$, initial state $q_0 = L_\varphi$, set of final states $F = \{\, q \in Q \mid \lambda \in q \,\}$, and where the transition functions $\cdot$ and $\circ$ are defined by $q \cdot a = a^{-1}q$ and $q \circ a = qa^{-1}$, for all $q \in Q$ and $a \in \Sigma$. Further let $k' = |Q| - k$. The fact that the instance (A_φ, k') can be constructed in polynomial time from the given instance (φ, k) of the MAX-2-SAT problem can be seen as follows. Every clause language induces a fixed number of states in A_φ, and except for the $\$$ symbol the clause languages are defined over disjoint alphabets. Hence the number of states corresponding to clause languages is linear in the number of clauses. Further, the number of states induced by the variable languages is $O(mn)$, so the automaton A_φ can be constructed in polynomial time.

Before we show the correctness of the reduction, let us analyze the structure of A_φ. The preamble of A_φ consists of the states

$$\mathrm{Pre}(A_\varphi) = \{q_0\} \cup \{\, q_0 \cdot b_{x,h}^{(i)}, q_0 \circ b_{x,h}^{(i)} \mid 1 \le i \le m, 1 \le h \le m+1, x \in \{1,2\} \,\},$$

which can be seen as follows. Let $1 \leq i \leq m$ and $1 \leq h \leq m + 1$, and consider clause $\varphi_i = (\ell_{i_1} \vee \ell_{i_2})$, with $\ell_{i_1} \in \{x_u, \overline{x}_u\}$ and $\ell_{i_1} \in \{x_v, \overline{x}_v\}$. Then

$$q_0 \cdot b_{1,h}^{(i)} = e_1^{(i)^+} + d^{(i)^*} c^{(i)}, \qquad\qquad q_0 \circ b_{1,h}^{(i)} = \#_{u,h} \$^* \#_{u,h},$$

$$q_0 \cdot b_{2,h}^{(i)} = e_2^{(i)^+} + d^{(i)^+} c^{(i)}, \qquad\qquad q_0 \circ b_{2,h}^{(i)} = \#_{v,h} \$^* \#_{v,h}.$$

One can verify from the descriptions of the languages L_{φ_i} and L_{x_j} that these states are only reachable by reading a single symbol from $\bigcup_{i=1}^m (B_1^{(i)} \cup B_2^{(i)})$, so they are preamble states. By examining all other states that can be reached from the initial state by reading a symbol which is not in $\bigcup_{i=1}^m (B_1^{(i)} \cup B_2^{(i)})$, one can see that there are no further preamble states. As an example, let us consider state $q_0 \cdot a_1^{(i)}$: this state can also be reached from q_0 by reading words from $\$^+ f_1$ with forward transitions and words from $f_1 \$^+$ with backward transitions. Therefore, state $q_0 \cdot a_1^{(i)}$ and all states reachable from it are kernel states. The reader is invited to verify that all other states of A_φ are kernel states, too.

Figure 10.3 on the next page, which is similar to Figure 10.2 shown before the proof, exemplarily shows a part of the structure of A_φ that corresponds to a clause language L_{φ_i}, for $\varphi_i = (x_u \vee \overline{x}_v)$. The states are renamed as follows:

$$q_1^{(i)} := q_0 \cdot a_1^{(i)} = e_1^{(i)^*} + d^{(i)^+} c^{(i)},$$

$$q_1'^{(i)} := q_0 \cdot b_{1,h}^{(i)} = e_1^{(i)^+} + d^{(i)^*} c^{(i)}, \text{ for } 1 \leq h \leq m+1,$$

$$q_2^{(i)} := q_0 \cdot a_2^{(i)} = e_2^{(i)^+} + d^{(i)^*} c^{(i)},$$

$$q_2'^{(i)} := q_0 \cdot b_{2,h}^{(i)} = e_2^{(i)^+} + d^{(i)^+} c^{(i)}, \text{ for } 1 \leq h \leq m+1,$$

$$s^{(i)} := q_0 \circ c^{(i)} = (a_1^{(i)} + B_2^{(i)}) \cdot d^{(i)^+} + (a_2^{(i)} + B_1^{(i)}) \cdot d^{(i)^*},$$

$$s_1^{(i)} := (q_0 \cdot \$f_4^{(i)}) \circ f_4^{(i)} \$ = (a_1^{(i)} + B_1^{(i)} + B_2^{(i)}) \cdot d^{(i)^+} + a_2^{(i)} \cdot d^{(i)^*},$$

$$s_2^{(i)} := (q_0 \cdot \$f_5^{(i)}) \circ f_5^{(i)} \$ = a_1^{(i)} \cdot d^{(i)^+} + (a_2^{(i)} + B_1^{(i)} + B_2^{(i)}) \cdot d^{(i)^*},$$

$$t_1^{(i)} := q_1^{(i)} \circ c^{(i)} = d^{(i)^+},$$

$$t_2^{(i)} := q_2^{(i)} \circ c^{(i)} = d^{(i)^*}.$$

Note that the following almost-equivalences hold between those states:

$$q_1^{(i)} \sim q_1'^{(i)}, \quad q_2^{(i)} \sim q_2'^{(i)}, \quad s^{(i)} \sim s_1^{(i)} \sim s_2^{(i)}, \quad t_1^{(i)} \sim t_2^{(i)}.$$

In fact, the mentioned states are not almost-equivalent to any other states, which can be seen as follows. The right languages of above described states for the clause φ_i contain infinitely many words with $d^{(i)}$ symbols. Therefore, these states cannot be almost-equivalent to states which correspond to derivatives of some clause language L_{φ_j}, with $j \neq i$, nor to states corresponding to derivatives of variable languages. Thus, the above states could only be almost-equivalent to

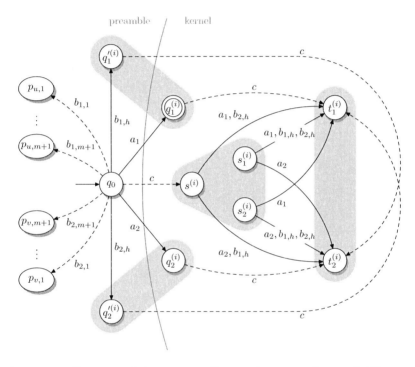

Figure 10.3: Structure of A_φ corresponding to clause $\varphi_i = (x_u \vee \overline{x}_v)$. The gray shading denotes almost-equivalence of states.

states that correspond to some derivative of the clause language L_{φ_i}. But this is also not the case, which can be verified by inspection of the regular expression for L_{φ_i}. Finally, notice that also the states

$$p_{u,h} := q_0 \circ b_{1,h}^{(i)} = \#_{u,h}\$^* \#_{u,h} \quad \text{and} \quad p_{v,h} := q_0 \circ b_{2,h}^{(i)} = \#_{v,h}\$^* \#_{v,h},$$

for $1 \leq h \leq m+1$, which correspond to the variables x_u and x_v, respectively, are not almost-equivalent to any other state.

Now we are ready to show the correctness of the reduction. To do this, we prove the following two claims—recall that Q is the state set of A_φ:

Claim (1). If there exists a biautomaton with $k' = |Q| - k$ states that is almost-equivalent to A_φ, then there exists a truth assignment that satisfies at least k clauses of φ.

Claim (2). If there is a truth assignment that satisfies at least k clauses of φ, then there exists a biautomaton with $k' = |Q| - k$ states that is almost-equivalent to A_φ.

Proof of Claim (1). Let $B = (Q', \Sigma, \cdot', \circ', q_0', F')$ be a minimal biautomaton with at most k' states, satisfying $B \sim A_\varphi$. Then we know from Theorem 10.2 that the kernels of the two automata are isomorphic. So in the following we may identify the kernel states of B with the derivatives of L_φ that constitute the kernel states of A_φ. Since $|Q'| \leq |Q| - k$, and the kernels of both automata are isomorphic, the preamble of B contains at most $|\text{Pre}(A_\varphi)| - k$ states. The only states in $\text{Pre}(A_\varphi)$ that can be saved are the states $q_1'^{(i)} = q_0 \cdot b_{1,h}^{(i)}$ and $q_2'^{(i)} = q_0 \cdot b_{2,h}^{(i)}$, with $1 \leq i \leq m$ and $1 \leq h \leq m+1$, because the other preamble states q_0 and $q_0 \circ b_{x,h}^{(i)}$, for $x \in \{1,2\}$ and $1 \leq i \leq m$, are not almost-equivalent to any other states. A state $q_1'^{(i)}$ (or $q_2'^{(i)}$) of A_φ can be saved if and only if in B we have $q_0' \cdot' b_{1,h}^{(i)} = q_0' \cdot' a_1^{(i)}$ (or $q_0' \cdot' b_{2,h}^{(i)} = q_0' \cdot' a_2^{(i)}$, respectively), for all h with $1 \leq h \leq m+1$, that is, if state $q_1'^{(i)}$ (or $q_2'^{(i)}$) is merged to the almost-equivalent kernel state $q_1^{(i)}$ (or $q_2^{(i)}$, respectively).

We first show that at most one state can be saved in each clause gadget. Consider the case where for some clause $\varphi_i = (\ell_{i_1} \vee \ell_{i_2})$, with $1 \leq i \leq m$, both $q_1'^{(i)}$ and $q_2'^{(i)}$ are merged to the kernel states $q_1^{(i)}$ and $q_2^{(i)}$, respectively. This means that in B the following holds for all h, with $1 \leq h \leq m+1$:

$$q_0' \cdot' b_{1,h}^{(i)} = \begin{cases} e_1^{(i)*} + d^{(i)+}c^{(i)} & \text{if } \ell_{i_1} \text{ is positive,} \\ e_1^{(i)+} + d^{(i)+}c^{(i)} & \text{if } \ell_{i_1} \text{ is negative,} \end{cases}$$

$$q_0' \cdot' b_{2,h}^{(i)} = \begin{cases} e_2^{(i)*} + d^{(i)*}c^{(i)} & \text{if } \ell_{i_2} \text{ is positive,} \\ e_2^{(i)+} + d^{(i)*}c^{(i)} & \text{if } \ell_{i_2} \text{ is negative.} \end{cases}$$

It follows $(q_0' \cdot' b_{1,h}^{(i)}) \circ' c^{(i)} = d^{(i)^+}$ and $(q_0' \cdot' b_{2,h}^{(i)}) \circ' c^{(i)} = d^{(i)^*}$, for $1 \le h \le m+1$. Since B has the $\diamond$-property, the state $s'^{(i)} := q_0' \circ c^{(i)}$ must lead to state $d^{(i)^+}$ on symbols $b_{1,h}^{(i)}$ and $a_1^{(i)}$, and it must lead to state $d^{(i)^*}$ on symbols $b_{2,h}^{(i)}$ and $a_2^{(i)}$. Such a state $s'^{(i)}$ is not present in automaton A_φ (though it is almost equivalent to the states $s^{(i)}$, $s_1^{(i)}$, and $s_2^{(i)}$), so it must be an additional preamble state of B.

Thus, for each clause φ_i, the number of states in $\mathrm{Pre}(B)$ that can be reached by reading a symbol from $\Sigma^{(i)}$ is equal to or by one smaller than the number of states in $\mathrm{Pre}(A_\varphi)$ reachable by such symbols. Since we have $|\mathrm{Pre}(B)| \le |\mathrm{Pre}(A_\varphi)| - k$, there must be k clauses $\varphi_{\alpha_1}, \varphi_{\alpha_2}, \ldots \varphi_{\alpha_k}$, such that for each number $\alpha \in \{\alpha_1, \alpha_2, \ldots, \alpha_k\}$ the state $q_1'^{(\alpha)}$ is merged to $q_1^{(\alpha)}$, or state $q_2'^{(\alpha)}$ is merged to $q_2^{(\alpha)}$. We use this information to construct the truth assignment $\chi \colon X \to \{0,1\}$ for φ as follows. For each index $\alpha \in \{\alpha_1, \alpha_2, \ldots, \alpha_k\}$ consider the clause $\varphi_\alpha = (\ell_{\alpha_1} \vee \ell_{\alpha_2})$.

- If $q_1'^{(\alpha)}$ is merged to $q_1^{(\alpha)}$ and $\ell_{\alpha_1} = x_u$, for some $x_u \in X$, then define $\chi(x_u) = 1$.

- If $q_1'^{(\alpha)}$ is merged to $q_1^{(\alpha)}$ and $\ell_{\alpha_1} = \overline{x}_u$, for some $x_u \in X$, then define $\chi(x_u) = 0$.

- If $q_2'^{(\alpha)}$ is merged to $q_2^{(\alpha)}$ and $\ell_{\alpha_2} = x_v$, for some $x_v \in X$, then define $\chi(x_v) = 1$.

- If $q_2'^{(\alpha)}$ is merged to $q_2^{(\alpha)}$ and $\ell_{\alpha_2} = \overline{x}_v$, for some $x_v \in X$, then define $\chi(x_v) = 0$.

For all remaining variables $x_j \in X$, which have not yet been assignment a truth value, we choose $\chi(x_j) = 0$. Let us first see that there is no variable which gets assigned both values 0 and 1. Assume that there is such a variable x_j. Then there must be clauses φ_α and $\varphi_{\alpha'}$ with $\alpha, \alpha' \in \{\alpha_1, \alpha_2, \ldots, \alpha_k\}$, such that the variable x_j appears positive in φ_α and negative in $\varphi_{\alpha'}$. Assume that $x_j = \ell_{\alpha_1}$ is the first literal of φ_α and $\overline{x}_j = \ell_{\alpha_2'}$ is the second literal of α'—the other cases can be handled similarly. Since x_j gets assigned truth value 1, the state $q_1'^{(\alpha)}$ must have been merged to $q_1^{(\alpha)}$, which means that for all h, with $1 \le h \le m+1$, we have

$$q_0' \cdot' b_{1,h}^{(\alpha)} = e_1^{(\alpha)^*} + d^{(\alpha)^+} c^{(\alpha)} \in F'.$$

Since B has the F-property, also the preamble states $q_0' \circ' b_{1,h}^{(\alpha)}$, for $1 \le h \le m+1$, must be in F'. Recall that the state $(q_0' \circ' b_{1,h}^{(\alpha)})$ of automaton B is almost-equivalent to state $q_0 \circ b_{1,h}^{(\alpha)} = p_{j,h}$ of A_φ, for $1 \le h \le m+1$. Since x_j also gets assigned truth value 0, the state $q_2'^{(\alpha')}$ must have been merged to $q_2^{(\alpha')}$. This means that for all h, with $1 \le h \le m+1$, we have

$$q_0' \cdot' b_{2,h}^{(\alpha')} = e_2^{(\alpha')^+} + d^{(\alpha')^*} c^{(\alpha')} \notin F'.$$

Since B has the F-property, also the preamble states $q_0' \circ' b_{2,h}^{(\alpha')}$, for $1 \leq h \leq m+1$, do not belong to F'. Notice that the state $q_0' \circ' b_{2,h}^{(\alpha')}$ is almost-equivalent to the state $q_0 \circ b_{2,h}^{(\alpha')} = p_{j,h}$ of A, for $1 \leq h \leq m+1$. It follows that for each state $p_{j,h}$ from $\mathrm{Pre}(A_\varphi)$, with $1 \leq h \leq m+1$, there are two different states $q_0' \circ b_{1,h}^{(\alpha)}$ and $q_0' \circ b_{2,h}^{(\alpha')}$ in $\mathrm{Pre}(B)$. But with these $m+1$ additional states, the preamble of B is definitely larger than the preamble of A_φ, since for each of the m clauses at most one state can be saved. This is a contradiction, so every variable gets assigned exactly one truth value.

From the definition of the truth assignment χ it clearly follows that χ satisfies each clause φ_α, for $\alpha \in \{\alpha_1, \alpha_2, \ldots, \alpha_k\}$. We have shown that if there is a biautomaton B, with $A_\varphi \sim B$, that has $k' = |Q| - k$ states, then there is a truth assignment χ that satisfies k clauses of φ. □

It remains to prove the second claim.

Proof of Claim (2). Let $\chi \colon X \to \{0,1\}$ be a truth assignment that satisfies the clauses $\varphi_{\alpha_1}, \varphi_{\alpha_2}, \ldots, \varphi_{\alpha_k}$. We construct the biautomaton B from automaton A_φ as follows. For all variables $x_j \in X$, with $\chi(x_j) = 1$, we make the states $p_{j,h}$ accepting, for $1 \leq h \leq m+1$. Note that $p_{j,h}$ is only reachable from q_0 by reading a symbol from $B_{x_j,h}$ with a backward transition. In order to preserve the F-property, we also make the states $q_0 \cdot b$ accepting, for all $b \in B_{x_j,h}$. Since all these states are in the preamble, we still have an almost-equivalent $\diamond$-F-DBiA.

Now let $i \in \{\alpha_1, \alpha_2, \ldots, \alpha_k\}$ and consider the clause $\varphi_i = (\ell_{i_1} \vee \ell_{i_2})$. If φ_i is satisfied by the first literal ℓ_{i_1}, then we can merge state $q_1'^{(i)}$ to state $q_1^{(i)}$ by re-routing all forward transitions from q_0 on symbols $b_{1,h}^{(i)}$ to state $q_1^{(i)}$, and re-routing the backward transition from q_0 on symbol $c^{(i)}$ to state s_1. Now the state $q_1'^{(i)}$ is not reachable anymore and can be removed. If the clause φ_i is not satisfied by literal ℓ_{i_1}, then it must be satisfied by the second literal ℓ_{i_2}. In this case, we can re-route all forward transitions from q_0 on symbols $b_{2,h}^{(i)}$ to state $q_2^{(i)}$ and, re-route the backward transition from q_0 on symbol $c^{(i)}$ to state s_2, so that state $q_2'^{(i)}$ can be removed. Note that the resulting biautomaton still has the $\diamond$-property and the F-property. Further, it is almost-equivalent to A_φ, because we only re-routed transitions starting from preamble states, and the new target states for these transitions are almost-equivalent to the original target states. In this way we can remove k states from A_φ and obtain an almost-equivalent $\diamond$-F-DBiA A' that has $k' = |Q| - k$ states. We have shown that if k clauses of φ can be satisfied, then k states of A_φ can be saved. □

The statement of the theorem now follows from Claims (1) and (2). □

Containment of the hyper-minimization problem in NP can be seen by an easy guess-and-check-algorithm. Therefore, we obtain the following theorem.

Theorem 10.6 (Hyper-Minimization Problem). *The problem of deciding for a given deterministic biautomaton A, that satisfies the $\diamond$-property and the F-property, and an integer n, whether there exists an almost-equivalent deterministic biautomaton B with n states, satisfying the $\diamond$- and F-property, is* NP-*complete.*

Proof. We know from Theorem 10.5 that the problem is NP-hard, hence it remains to prove containment in NP. Given as an instance of the problem a biautomaton A and an integer n, a non-deterministic Turing machine first guesses a $\diamond$-F-DBiA B with n states and writes it on its working tape. Checking whether A and B are almost-equivalent can be done by testing whether the contained DFAs A' and B', obtained from A and B by considering only forward transitions, are almost-equivalent. It is shown in Section 4.1 that this question for DFAs can be decided in NL. Therefore, the hyper-minimization problem for biautomata can be decided by a non-deterministic Turing machine in polynomial time. □

In the remaining part of this section we discuss the E-minimization problem for biautomata. We recall the following definitions from Section 3.2. Let Σ be an alphabet and $E \subseteq \Sigma^*$ be a language, called the *error language*. Two languages L and L' over Σ are *E-equivalent* ($L \sim_E L'$) if $L \triangle L' \subseteq E$. Now two biautomata A and A' are *E-equivalent* ($A \sim_E A'$) if $L(A) \sim_E L(A')$. We say that a biautomaton A is *E-minimal* if there is no biautomaton A', with $A \sim_E A'$, that has fewer states than A.

We have seen in Section 4.2 that the E-minimization problem for DFAs is NP-complete. In the following we show that the same holds for the corresponding problem for biautomata. Concerning the definition of the E-minimization problem for biautomata, one could consider two possibilities for specifying the error set E: either by a deterministic finite automaton, or by a biautomaton. The next result shows that both variants turn out to be NP-complete.

Theorem 10.7 (E-Minimization Problem). *The problem of deciding for a given deterministic biautomaton A, that satisfies the $\diamond$-property and the F-property, a deterministic finite automaton A_E, and an integer n, whether there exists a deterministic biautomaton B with n states, satisfying the $\diamond$- and F-property, such that $A \sim_E B$, for $E = L(A_E)$, is* NP-*complete. This statement also holds, if A_E is a deterministic biautomaton, satisfying the $\diamond$-property and the F-property, instead of a finite automaton.*

Proof. For the upper bound we can use a simple guess-and-check algorithm. Given the automata A and A_E, and an integer n, a nondeterministic Turing machine can guess an n-state biautomaton B and then simply has to check whether $A \sim_E B$. For this the machine can use the NL algorithm from Section 3.1 to check E-equivalence of the DFAs A' and B', which are obtained from A and B, respectively, by considering only forward transitions. If the automaton A_E is a biautomaton instead of a finite automaton, then the DFA A'_E, which is obtained similarly from A_E, is used instead of A_E. This shows that the problem can be solved by a nondeterministic Turing machine in polynomial time.

The lower bound of NP-hardness follows by a simple reduction from the hyper-minimization problem for biautomata, which was shown NP-complete in Theorem 10.6. Let the biautomaton A, with input alphabet Σ, and integer n form an instance of the hyper-minimization problem. Further let n' be the number of states of A, and $m = \max(n, n')$. We choose the error language $E = \Sigma^{\leq m-2}$. An automaton A_E with $L(A_E) = E$ can easily be constructed from the numbers n' and m, no matter whether it needs to be a DFA or a biautomaton. Now the instance to the E-minimization problem consists of the automata A and A_E, and of the integer n. It remains to explain why there exists an n-state biautomaton B with $A \sim_E B$ if and only if there exists an n-state biautomaton B that satisfies $A \sim B$. This follows from Theorem 3.12 from Part II, which says that if two languages L and L' are accepted by two DFAs having n and n' states, respectively, then $L \sim L$ if and only if $L \sim_E L'$, for $E = \Sigma^{\leq m-2}$, where $m = \max(n, n')$. Since any biautomaton, with $\diamond$- and F-property, contains an equivalent DFA, the correctness of the reduction follows. $\qquad\qquad\qquad\qquad\qquad\qquad\qquad\qquad\qquad\qquad\qquad\qquad\quad\square$

Table 10.1 compares the complexities of the different minimization problems for deterministic biautomata and for deterministic finite automata with respect to different error profiles.

Minimization Problem		
Equivalence relation	DFA	$\diamond$-F-DBiA
$\equiv$	NL	
$\sim$	NL	NP
$\sim_E$		NP

Table 10.1: Computational complexity of minimizing deterministic finite automata and deterministic biautomata, that satisfy the $\diamond$- and F-property, with respect to different equivalence relations. An entry indicates that the corresponding problem is complete for the stated complexity class.

Part IV

Outro

Chapter 11

Conclusions and Directions for Further Research

THIS final chapter closes the thesis with a recapitulation of our main findings as well as some proposals for further research. Our investigations of modern aspects in the theory of automata and regular languages produced results in the fields of computational and descriptional complexity theory, in language theory, as well as results on algorithms for automaton minimization.

Concerning computational complexity, we focused on problems related to automata minimization. In particular, Part II was devoted to lossy compression of finite automata. We introduced the notions of E-equivalence and E-minimality as generalizations of different equivalence and minimization concepts from the literature, such as classical equivalence and minimization, almost-equivalence and hyper-minimality, and cover languages and minimal cover automata. These equivalence concepts served as error profiles for lossy compression of automata. In Chapter 4 we have seen that, in terms of computational complexity, hyper-minimization of deterministic finite automata is as easy as classical minimization, namely NL-complete. On the other hand, we proved that the general E-minimization problem is computationally intractable, namely NP-complete, already for DFAs. This increase in the complexity, when changing from almost-equivalence to E-equivalence, was also observed when considering counting problems for minimal automata. The main computational complexity results in Part III on biautomata were the following. We showed that the hyper-minimization problem for deterministic biautomata, that satisfy the $\diamond$- and the F-property, is already NP-complete. This is a remarkable difference between finite automata and biautomata with $\diamond$- and F-property, because these two automaton models are in many ways quite similar. For example, it expectedly turned out that exact minimization of $\diamond$-F-DBiAs is NL-complete—just as it is for DFAs.

Coming to algorithms for automata minimization, we showed in Chapter 9 how classical minimization algorithms for finite automata can successfully be applied to biautomata. Here the $\diamond$- and F-properties turned out to be essential for efficient minimization: for biautomata without the $\diamond$-property there is no minimization algorithm at all. In presence of both properties also the rather inconvenient minimization technique of Brzozowski, that uses the power-set and the dual construction twice, works correctly for biautomata. However, the algorithm behaves a bit strange concerning the structure and accepted language of the intermediate dual automaton—see Section 7.3. As we have seen in Chapter 5 at the end of Part II, Brzozowski's minimization algorithm can also be adapted so that it can be used for lossy compression of finite automata. Here the treatment of different equivalence relations between states in terms of the more general concept of E-equivalence gave a useful basis for the development of the different algorithms.

Some results in language theory were obtained in Chapter 7 of Part III. Here the influence of the three investigated structural properties for biautomata—the $\diamond$-, F-, and I-properties—on the accepted languages were studied. It turned out that the F-property alone does not change the class of accepted languages. The general biautomaton model characterizes the class of linear context-free languages, while biautomata, that satisfy the $\diamond$-property, accept regular languages. The I-property, together with the $\diamond$-property yields the class of cyclic regular languages, and replacing the I-property by an even stronger restriction yields exactly the commutative regular languages.

Finally, recall the descriptional complexity results from Chapter 8. We studied conversions from classical descriptional systems—deterministic and nondeterministic finite automata, regular expressions, and syntactic monoids—to biautomata that satisfy the $\diamond$-property and the F-property. Mostly tight bounds for these conversions were obtained, where an adaptation of the fooling set technique for finite automata to biautomata proved useful for obtaining lower bounds. Concerning conversions between different types of biautomata, we showed that determinization of nondeterministic biautomata induces an exponential blow-up in the number of states. Enforcing biautomata to satisfy the F-property is cheap, from a descriptional complexity point of view, if we do not require to keep the $\diamond$-property. But if the $\diamond$-property has to be preserved, then enforcing the F-property causes an exponential size increase for deterministic biautomata. Moreover, from the fact that the $\diamond$-property marks the border between the regular languages and the linear context-free languages, we obtained a non-recursive trade-off between biautomata in general, and those which satisfy the $\diamond$-property.

Having reviewed our main results, it is now time to stimulate further research on the central topics of this thesis, namely lossy compression of language representations and the theory of biautomata. Problems addressed but not fully solved in this work, like for example non-matching upper and lower bounds in descriptional or computational complexity, were already mentioned in the corresponding sections. Here we do not want to repeat these, but rather tend to highlight some new directions for further research.

We have characterized the complexities of various problems related to automata that are minimal with respect to almost-equivalence and to E-equivalence. A similar study could also be undertaken for cover automata, both in the finite automaton model, as well as for biautomata. Because of the different complexities for hyper-minimization of finite automata and biautomata, the complexity of finding minimal deterministic cover biautomata is of particular interest. Moreover, when faced with computationally intractable minimization problems, one could ask whether the minimal automaton can at least be approximated. In particular the approximation problem for hyper-minimal biautomata seems worth looking at.

The different complexities of the minimization problems also lead us to another related line of research: efficient minimization methods often rely on structural conditions on the automata. An aspect, that goes hand in hand with the efficient hyper-minimization of DFAs and the non-efficient hyper-minimization of DBiAs, is the rich structural similarity between almost-equivalent DFAs and the poorer similarity between almost-equivalent DBiAs. From this point of view it seems an important task to further study the structure of automata that are equivalent with respect to certain error profiles—here cover automata seem a suitable object for further study. For example, is there a meaningful definition of kernel parts of cover automata that allows to find isomorphisms between the kernels of equivalent cover automata?

Another challenging goal is to extend the tractability frontier of lossy compression beyond hyper-minimization and minimal cover automata. The error sets associated to these error profiles are of very special form, namely $E = \Sigma^{\leq \ell}$ and $E = \Sigma^{\geq \ell}$, respectively, for appropriate integers ℓ. Maybe one can find suitable conditions on the error set E that are on the one hand strong enough to make E-minimization practically solvable, but which on the other hand are flexible enough to allow more control on the committed errors than in the case of almost-equivalence and cover automata. As a possible first step in this direction, one could investigate which properties allow minimization by state merging methods.

Generally, besides for the herein studied automaton models, it may be interesting to consider lossy compression also for other forms of language representations, like for example regular expressions or formalism for non-regular languages.

Concerning research on biautomata, we suggest to further investigate descriptional complexity aspects of this automaton model. A first question concerns the opposite direction of the conversions studied in this work: what can be said about the descriptional complexity of converting biautomata to other models? Biautomata that do not even satisfy the ◇-property induce a nonrecursive trade-off in comparison to other descriptional models for regular languages. Therefore, we concentrate on biautomata with the ◇-property. Since a biautomaton that satisfies the ◇-property and the F-property contains an equivalent finite automaton, we can inherit the upper bounds for the corresponding conversions from finite automata. However, these bounds may not be best possible in general. Moreover, similar conversion problems can be considered for biautomata that satisfy ◇-property but not necessarily the F-property. Such a biautomaton can be converted into an equivalent nondeterministic finite automaton with a quadratic increase of the number of states. This can be combined with known complexity bounds for conversions from NFAs to other models to yield upper bounds for corresponding biautomaton conversions.

Besides conversion problems, also the complexity of applying language operations is a typical research topic in descriptional complexity theory. For example, given an n-state automaton accepting some regular language, how many states are needed in an automaton for the reversal of that language? While the reversal operation is expensive for deterministic finite automata—basically one has to build the power-set automaton of the reversal automaton—it is easily implemented on a biautomaton: simply interchange the forward and backward transition functions. A more challenging problem is to build a biautomaton that accepts the concatenation of the languages of two given biautomata. Let us consider the case where the ◇- and F-properties are required. Then the classical approach of connecting the accepting states of the first automaton to the initial state of the second automaton does not work here, because this construction can destroy the ◇-property. A fail-safe construction would be to convert the given biautomata into finite automata—in presence of ◇- and F-property this comes for free—then implement concatenation on finite automata, and finally convert the finite automaton into a biautomaton. For nondeterministic machines, this approach is not too expensive. But in case of deterministic automata, the finite automaton construction for concatenation induces an exponential size blow-up and also the conversion from DFAs to deterministic biautomata is of exponential cost. So in combination this yields a double-exponential upper bound for the cost of concatenation of biautomaton languages. The question is whether this is really best possible.

Finally, we like to draw the attention to unambiguous automata, which are nondeterministic machines where for every word there is at most one accepting computation. This can be seen as a working mode between determinism and nondeterminism. In fact, exponential size trade-offs between nondeterministic and unambiguous finite automata, as well as between unambiguous and deterministic finite automata are known. While this notion is meaningful for finite automata, it is not immediately clear how to define unambiguous biautomata. The idea that a biautomaton should admit at most one accepting computation for each word conflicts with the inherent ambiguity of biautomata with ◇- and F-property: acceptance of a word by such a biautomaton is independent from the order of the head movement and the position in the word, where the heads meet. A possible solution might be to say that there is at most one accepting computation for each partitioning of a word and each corresponding order of head movements.

Bibliography

[1] Alfred V. Aho, John E. Hopcroft, and Jeffrey D. Ullman. *The Design and Analysis of Computer Algorithms*. Addison-Wesley, 1974.

[2] Marco Almeida, Nelma Moreira, and Rogério Reis. On the performance of automata minimization algorithms. In Arnold Beckman, Costas Dimitracopoulos, and Benedikt Löwe, editors, *Proceedings of the 4th Conference on Computability in Europe—Logic and Theory of Algorithms*, pages 3–14, Athens, Greece, June 2008. University of Athens.

[3] Carme Àlvarez and Birgit Jenner. A very hard log-space counting class. *Theoretical Computer Science*, 107(1):3–30, January 1993.

[4] Andrew Badr. Hyper-minimization in $O(n^2)$. *International Journal of Foundations of Computer Science*, 20(4):735–746, August 2009.

[5] Andrew Badr, Viliam Geffert, and Ian Shipman. Hyper-minimizing minimized deterministic finite state automata. *RAIRO - Theoretical Informatics and Applications*, 43(1):69–94, January 2009.

[6] Christel Baier and Joost-Pieter Katoen. *Priciples of Model Checking*. MIT Press, 2008.

[7] Gerard Berry and Ravi Sethi. From regular expressions to deterministic automata. *Theoretical Computer Science*, 48:117–126, December 1986.

[8] Jean-Camille Birget. Intersection and union of regular languages and state complexity. *Information Processing Letters*, 43(4):185–190, September 1992.

[9] Henrik Björklund and Wim Martens. The tractability frontier for NFA minimization. *Journal of Computer and System Sciences*, 78(1):198–210, January 2012.

[10] Filippo Bonchi and Damien Pous. Checking NFA equivalence with bisimulations up to congruence. In Roberto Giacobazzi and Radhia Cousot,

editors, *Proceedings of the 40th Annual ACM SIGPLAN-SIGACT Symposium on Principles of Programming Languages*, pages 457–468, Rome, Italy, January 2013. ACM.

[11] Raymond Boute and Edward J. McCluskey. Fault equivalence in sequential machines. In Jerome Fox, editor, *Proceedings of the Symposium on Computers and Automata*, volume 21 of *Microwave Research Institute Symposia Series*, pages 483–507, New York, USA, April 1971. Polytechnic Press of the Polytechnic Institute of Brooklyn.

[12] Janusz A. Brzozowski. Canonical regular expressions and minimal state graphs for definite events. In Jerome Fox, editor, *Proceedings of the Symposium on Mathematical Theory of Automata*, volume 12 of *MRI Symposia Series*, pages 529–561, New York, USA, April 1962. Polytechnic Press of the Polytechnic Institute of Brooklyn.

[13] Janusz A. Brzozowski. Derivatives of regular expressions. *Journal of the Association for Computing Machinery*, 11(4):481–494, October 1964.

[14] Janusz A. Brzozowski and Edward J. McCluskey. Signal flow graph techniques for sequential circuit state diagrams. *IEEE Transactions on Electronic Computers*, EC-12(2):67–76, April 1963.

[15] J. Richard Büchi. Weak second-order arithmetic and finite automata. *Zeitschrift für Mathematische Logik und Grundlagen der Mathematik*, 6 (1–6):66–92, 1960.

[16] Cezar Câmpeanu. Simplifying nondeterministic finite cover automata. In Ésik and Fülöp [31], pages 162–173.

[17] Cezar Câmpeanu, Nicolae Sântean, and Sheng Yu. Minimal cover-automata for finite languages. *Theoretical Computer Science*, 267(1–2): 3–16, September 2001.

[18] Cezar Câmpeanu, Andrei Păun, and Sheng Yu. An efficient algorithm for constructing minimal cover automata for finite languages. *International Journal of Foundations of Computer Science*, 13(1):83–97, February 2002.

[19] Cezar Câmpeanu, Andrei Păun, and Jason R. Smith. Incremental construction of minimal deterministic finite cover automata. *Theoretical Computer Science*, 363(2):135–148, October 2006.

[20] Jean-Marc Champarnaud, Ahmed Khorsi, and Thomas Paranthoën. Split and join for minimizing: Brzozowski's algorithm. In Miroslav Balík and Milan Šimánek, editors, *Proceedings of the Prague Stringology Conference*, number DC–2002–03 in Research Report, pages 96–104, Prague, Czech Republic, September 2002. Czech Technical University.

[21] Jean-Marc Champarnaud, Franck Guingne, and Georges Hansel. Similarity relations and cover automata. *RAIRO - Theoretical Informatics and Applications*, 39(1):115–123, March 2005.

[22] Sang Cho and Dung T. Huynh. The parallel complexity of finite-state automata problems. *Information and Computation*, 97(1):1–22, March 1992.

[23] Noam Chomsky. Three models for the description of language. *IRE Transactions on Information Theory*, 2(3):113–124, September 1956.

[24] Noam Chomsky. On certain formal properties of grammars. *Information and Control*, 2(2):137–167, June 1959.

[25] Noam Chomsky. Context-free grammars and pushdown storage. Technical Report Quarterly Progress Report No. 65, MIT Research Laboratory of Electronics, Princeton, USA, April 1962.

[26] Noam Chomsky and George A. Miller. Finite state languages. *Information and Control*, 1(2):91–112, May 1958.

[27] Stephen A. Cook. The complexity of theorem-proving procedures. In Michael A. Harrison, Ranan B. Banerji, and Jeffrey D. Ullman, editors, *Proceedings of the 3rd Annual ACM Symposium on Theory of Computing*, pages 151–158, Shaker Heights, Ohio, USA, May 1971. ACM.

[28] Andrzej Ehrenfeucht and Paul Zeiger. Complexity measures for regular expressions. *Journal of Computer and System Sciences*, 12(2):134–146, April 1976.

[29] Calvin C. Elgot. Decision problems of finite automata design and related arithmetics. *Transactions of the American Mathematical Society*, 98(1): 21–51, January 1961.

[30] Keith Ellul, Bryan Krawetz, Jeffrey Shallit, and Ming-Wei Wang. Regular expressions: New results and open problems. *Journal of Automata, Languages and Combinatorics*, 9(2/3):233–256, September 2004.

[31] Zoltán Ésik and Zoltán Fülöp, editors. *Automata and Formal Languages*, volume 151 of *Electronic Proceedings in Theoretical Computer Science*, Szeged, Hungary, May 2014.

[32] R. James Evey. Application of pushdown-store machines. In James D. Tupac, editor, *AFIPS Conference Proceedings of the 1963 Fall Joint Computer Conference*, volume 24 of *Joint Computer Conference Proceedings*, pages 215–227, Las Vegas, USA, November 1963. Spartan Books.

[33] Michael R. Garey and David S. Johnson. *Computers and Intractability: A Guide to the Theory of NP-completeness*. W. H. Freeman, 1979.

[34] Paweł Gawrychowski and Artur Jeż. Hyper-minimisation made efficient. In Ratislav Královič and Damian Niwiński, editors, *Proceedings of the 34th International Symposium on Mathematical Foundations of Computer Science*, volume 5734 of *Lecture Notes in Computer Science*, pages 356–368, Novy Smokovec, High Tatras, Slovakia, August 2009. Springer.

[35] Paweł Gawrychowski, Artur Jeż, and Andreas Maletti. On minimising automata with errors. In Filip Murlak and Piotr Sankowski, editors, *Proceedings of the 36th International Symposium on Mathematical Foundations of Computer Science*, volume 6907 of *Lecture Notes in Computer Science*, pages 327–338, Warsaw, Poland, August 2011. Springer.

[36] Viliam Geffert, Carlo Mereghetti, and Giovanni Pighizzini. Converting two-way nondeterministic unary automata into simpler automata. *Theoretical Computer Science*, 295(1–3):189–203, February 2003.

[37] Wouter Gelade and Frank Neven. Succinctness of the complement and intersection of regular expressions. *ACM Transactions on Computational Logic*, 13(1):4:1–4:19, January 2012.

[38] Seymour Ginsburg and Sheila Greibach. Deterministic context free languages. *Information and Control*, 9(6):620–648, December 1966.

[39] Seymour Ginsburg and H. Gordon Rice. Two families of languages related to ALGOL. *Journal of the Association for Computing Machinery*, 9(3):350–371, July 1962.

[40] Viktor M. Glushkov. The abstract theory of automata. *Russian Mathematical Surveys*, 16(5):1–53, September–October 1961.

[41] E. Mark Gold. Complexity of automaton identification from given data. *Information and Control*, 37(3):302–320, June 1978.

[42] Hermann Gruber and Markus Holzer. Finite automata, digraph connectivity, and regular expression size. In Luca Aceto, Ivan Damgård, Leslie A. Goldberg, Magnús M. Halldórsson, Anna Ingólfsdóttir, and Igor Walukiewicz, editors, *Proceedings of the 35th International Colloquium on Automata, Languages and Programming, Part II*, volume 5126 of *Lecture Notes in Computer Science*, pages 39–50, Reykjavik, Iceland, July 2008. Springer.

[43] Hermann Gruber, Markus Holzer, and Martin Kutrib. On measuring non-recursive trade-offs. *Journal of Automata, Languages and Combinatorics*, 15(1/2):107–120, 2010.

[44] Leonard H. Haines. *Generation and Recognition of Formal Languages*. PhD Thesis, Massachusetts Institute of Technology, Deparment of Mathematics, Cambridge, Massachusetts, USA, June 1965.

[45] Michael A. Harrison. *Introduction to Formal Language Theory*. Addison-Wesley, 1978.

[46] Juris Hartmanis. On the succinctness of different representations of languages. *SIAM Journal on Computing*, 9(1):114–120, February 1980.

[47] Lane A. Hemaspaandra and Heribert Vollmer. The satanic notations: Counting classes beyond $\#P$ and other definitional adventures. *ACM SIGACT News*, 26(1):2–13, March 1995.

[48] Markus Holzer. On emptiness and counting for alternating finite automata. In Jürgen Dassow, Grzegorz Rozenberg, and Arto Salomaa, editors, *Developments in Language Theory II—At the Crossroads of Mathematics, Computer Science and Biology*, pages 88–97, Magdeburg, Germany, July 1995. World Scientific.

[49] Markus Holzer and Sebastian Jakobi. From equivalence to almost-equivalence, and beyond—minimizing automata with errors (extended abstract). In Yen and Ibarra [125], pages 190–201.

[50] Markus Holzer and Sebastian Jakobi. Brzozowski's minimization algorithm—more robust than expected (extended abstract). In Stavros Konstantinidis, editor, *Proceedings of the 18th International Conference on Implementation and Application of Automata*, volume 7982 of *Lecture Notes in Computer Science*, pages 181–192, Halifax, Nova Scotia, Canada, July 2013. Springer.

[51] Markus Holzer and Sebastian Jakobi. Nondeterministic biautomata and their descriptional complexity. In Helmut Jürgensen and Rogério Reis, editors, *Proceedings of the 15th International Workshop on Descriptional Complexity of Formal Systems*, volume 8031 of *Lecture Notes in Computer Science*, pages 112–123, London, Ontario, Canada, July 2013. Springer.

[52] Markus Holzer and Sebastian Jakobi. Minimization and characterizations for biautomata. In Suna Bensch, Frank Drewes, Rudolf Freund, and Friedrich Otto, editors, *Proceedings of the 5th Workshop on Non-Classical Models of Automata and Applications*, volume 294 of *books@ocg.at*, pages 179–193, Umeå, Sweden, August 2013. Austrian Computer Society.

[53] Markus Holzer and Sebastian Jakobi. From equivalence to almost-equivalence, and beyond: Minimizing automata with errors. *International Journal of Foundations of Computer Science*, 24(7):1083–1097, November 2013.

[54] Markus Holzer and Sebastian Jakobi. More structural characterizations of some subregular language families by biautomata. In Ésik and Fülöp [31], pages 271–285.

[55] Markus Holzer and Sebastian Jakobi. Minimal and hyper-minimal biautomata. In Arseny M. Shur and Mikhail V. Volkov, editors, *Proceedings of the 18th International Conference on Developments in Language Theory*, volume 8633 of *Lecture Notes in Computer Science*, pages 291–302, Ekaterinburg, Russia, August 2014. Springer.

[56] Markus Holzer and Sebastian Jakobi. Nondeterministic biautomata and their descriptional complexity. *International Journal of Foundations of Computer Science*, 24(7):837–855, November 2014.

[57] Markus Holzer and Sebastian Jakobi. Minimization and characterizations for biautomata. *Fundamenta Informaticae*, 136(1–2):113–137, January 2015.

[58] Markus Holzer and Martin Kutrib. Descriptional and computational complexity of finite automata—a survey. *Information and Computation*, 209 (3):456–470, March 2011.

[59] Markus Holzer and Andreas Maletti. An $n \log n$ algorithm for hyperminimizing a (minimized) deterministic automaton. *Theoretical Computer Science*, 411(38–39):3404–3413, August 2010.

[60] Markus Holzer, Martin Kutrib, and Andreas Malcher. Complexity of multi-head finite automata: Origins and directions. *Theoretical Computer Science*, 412(1–2):83–96, January 2011.

[61] John Hopcroft. An $n \log n$ algorithm for minimizing states in a finite automaton. In Zvi Kohavi and Azaria Paz, editors, *Proceedings of the International Symposium on Theory of Machines and Computations*, pages 189–196, Haifa, Israel, August 1971. Academic Press.

[62] John E. Hopcroft, Rajeev Motwani, and Jeffrey D. Ullman. *Introduction to Automata Theory, Languages, and Computation*. Addison-Wesley, third edition, 2006.

[63] David A. Huffman. The synthesis of sequential switching circuits, part I. *Journal of the Franklin Institute*, 257(3):161–190, March 1954.

[64] David A. Huffman. The synthesis of sequential switching circuits, part II. *Journal of the Franklin Institute*, 257(4):275–303, April 1954.

[65] Neil Immerman. Nondeterministic space is closed under complementation. *SIAM Journal on Computing*, 17(5):935–938, October 1988.

[66] Artur Jeż and Andreas Maletti. Hyper-minimization for deterministic tree automata. *International Journal of Foundations of Computer Science*, 24 (6):815–830, September 2013.

[67] Tao Jiang and Bala Ravikumar. Minimal NFA problems are hard. *SIAM Journal on Computing*, 22(6):1117–1141, December 1993.

[68] Galina Jirásková and Ondřej Klíma. Descriptional complexity of biautomata. In Martin Kutrib, Nelma Moreira, and Rogério Reis, editors, *Proceedings of the 14th International Workshop on Descriptional Complexity of Formal Systems*, volume 7386 of *Lecture Notes in Computer Science*, pages 196–208, Braga, Portugal, July 2012. Springer.

[69] Neil D. Jones. Space-bounded reducibility among combinatorial problems. *Journal of Computer and System Sciences*, 11(1):68–85, August 1975.

[70] Neil D. Jones, Y. Edmund Lien, and William T. Laaser. New problems complete for nondeterministic log space. *Mathematical Systems Theory*, 10(1):1–17, December 1976.

[71] Tsunehiko Kameda and Peter Weiner. On the state minimization of nondeterministic finite automata. *IEEE Transactions on Computers*, C-19 (7):617–627, July 1970.

[72] Richard M. Karp. Reducibility among combinatorial problems. In Raymond E. Miller, James W. Thatcher, and Jean D. Bohlinger, editors, *Complexity of Computer Computations*, IBM Research Symposia Series, pages 85–103, Yorktown Heights, New York, USA, May 1972. Plenum Press.

[73] Stephen C. Kleene. Representation of events in nerve nets and finite automata. In Claude E. Shannon and John McCarthy, editors, *Automata Studies*, volume AM-34 of *Annals of Mathematics Studies*, pages 3–41. Princeton University Press, Princeton, USA, April 1956.

[74] Ondřej Klíma and Libor Polák. On biautomata. *RAIRO - Theoretical Informatics and Applications*, 46(4):537–592, October 2012.

[75] Ondřej Klíma and Libor Polák. Biautomata for k-piecewise testable languages. In Yen and Ibarra [125], pages 344–355.

[76] Heiko Körner. A time and space efficient algorithm for minimizing cover automata for finite languages. *International Journal of Foundations of Computer Science*, 14(6):1071–1086, December 2003.

[77] Dexter Kozen. Lower bounds for natural proof systems. In *Proceedings of the 18th Annual Symposium on Foundations of Computer Science*, pages 254–266, Providence, Rhode Island, USA, October 1977. IEEE Computer Society.

[78] Sige-Yuki Kuroda. Classes of languages and linear-bounded automata. *Information and Control*, 7(2):207–223, June 1964.

[79] Martin Kutrib. The phenomenon of non-recursive trade-offs. *International Journal of Foundations of Computer Science*, 16(5):957–973, October 2005.

[80] Richard E. Ladner. Polynomial space counting problems. *SIAM Journal on Computing*, 18(6):1087–1097, December 1989.

[81] Ernst Leiss. The complexity of restricted regular expressions and the synthesis problem for finite automata. *Journal of Computer and System Sciences*, 23(3):348–354, December 1981.

[82] Hing Leung. Separating exponentially ambiguous finite automata from polynomially ambiguous finite automata. *SIAM Journal on Computing*, 27(4):1073–1082, August 1998.

[83] Leonid A. Levin. Universal sequential search problems. *Problemy Peredachi Informatsii*, 9(3):115–116, July–September 1973. (in Russian, English translation in [119]).

[84] Roussanka Loukanova. Linear context free languages. In Cliff B. Jones, Zhiming Liu, and Jim Woodcock, editors, *Proceedings of the 4th International Colloquium on Theoretical Aspects of Computing*, volume 4711 of *Lecture Notes in Computer Science*, pages 351–365, Macao, China, September 2007. Springer.

[85] Oleg B. Lupanov. A comparison of two types of finite sources. *Problemy Kibernetiki*, 9:321–326, 1963. (in Russian).

[86] Andreas Malcher. Minimizing finite automata is computationally hard. *Theoretical Computer Science*, 327(3):375–390, November 2004.

[87] Andreas Malcher. On recursive and non-recursive trade-offs between finite-turn pushdown automata. *Journal of Automata, Languages and Combinatorics*, 12(1/2):265–277, 2007.

[88] Andreas Maletti and Daniel Quernheim. Optimal hyper-minimization. *International Journal of Foundations of Computer Science*, 22(8):1877–1891, December 2011.

[89] Andreas Maletti and Daniel Quernheim. Unweighted and weighted hyper-minimization. *International Journal of Foundations of Computer Science*, 23(6):1207–1225, September 2012.

[90] Warren S. McCulloch and Walter Pitts. A logical calculus of the ideas immanent in nervous activity. *Bulletin of Mathematical Biophysics*, 5(4):115–133, December 1943.

[91] Robert McNaughton. Parenthesis grammars. *Journal of the Association for Computing Machinery*, 14(3):490–500, July 1967.

[92] Robert McNaughton and Hisao Yamada. Regular expressions and state graphs for automata. *IRE Transactions on Electronic Computers*, EC-9 (1):39–47, March 1960.

[93] George H. Mealy. A method for synthesizing sequential circuits. *The Bell Systems Technical Journal*, 34(5):1045–1079, September 1955.

[94] Albert R. Meyer and Michael J. Fischer. Economy of description by automata, grammars, and formal systems. In *Proceedings of the 12th Annual Symposium on Switching and Automata Theory*, pages 188–191, East Lansing, Michigan, USA, October 1971. IEEE Computer Society.

[95] Albert R. Meyer and Larry J. Stockmeyer. The equivalence problem for regular expressions with squaring requires exponential space. In *Proceedings of the 13th Annual Symposium on Switching and Automata Theory*, pages 125–129, College Park, Maryland, USA, October 1972. IEEE Computer Society.

[96] Edward F. Moore. Gedanken-experiments on sequential machines. In Claude E. Shannon and John McCarthy, editors, *Automata Studies*, volume AM-34 of *Annals of Mathematics Studies*, pages 129–153. Princeton University Press, Princeton, USA, April 1956.

[97] Frank R. Moore. On the bounds for state-set size in the proofs of equivalence between deterministic, nondeterministic, and two-way finite automata. *IEEE Transactions on Computers*, C-20(10):1211–1214, October 1971.

[98] John R. Myhill. Finite automata and the representation of events. Technical Report WADD TR-57-624, Wright Patterson Air Force Base, Ohio, USA, 1957.

[99] John R. Myhill. Linearly bounded automata. Technical Report WADD TR-60-165, Wright Patterson Air Force Base, Ohio, USA, 1960.

[100] Anil Nerode. Linear automaton transformations. *Proceedings of the American Mathematical Society*, 9(4):541–544, August 1958.

[101] Anthony G. Oettinger. Automatic syntactic analysis and the pushdown store. In Roman Jakobson, editor, *Structure of Language and its Mathematical Aspects*, volume 12 of *Proceedings of Symposia in Applied Mathematics*, pages 104–129, Providence, Rhode Island, USA, 1961. American Mathematical Society.

[102] Christos H. Papadimitriou. *Computational Complexity*. Addison-Wesley, 1994.

[103] Marvin C. Paull and Stephen H. Unger. Structural equivalence of context-free grammars. *Journal of Computer and System Sciences*, 2(4):427–463, December 1968.

[104] Jean-Éric Pin. Syntactic semigroups. In Rozenberg and Salomaa [107], chapter 10, pages 679–746.

[105] Michael O. Rabin and Dana Scott. Finite automata and their decision problems. *IBM Journal of Research and Development*, 3(2):114–125, April 1959.

[106] Arnold L. Rosenberg. On multi-head finite automata. *IBM Journal of Research and Development*, 10(5):388–394, September 1966.

[107] Grzegor Rozenberg and Arto Salomaa, editors. *Handbook of Formal Languages—Volume 1: Word, Language, Grammar*. Springer, 1997.

[108] Davide Sangiorgi. On the origins of bisimulation and coinduction. *ACM Transactions on Programming Languages and Systems*, 31(4):15:1–15:42, May 2009.

[109] Walter J. Savitch. Relationships between nondeterministic and deterministic tape complexities. *Journal of Computer and System Sciences*, 4(2): 177–192, April 1970.

[110] Sven Schewe. Beyond hyper-minimisation—minimising DBAs and DPAs is NP-complete. In Kamal Lodaya and Meena Mahajan, editors, *Proceedings of the 30th International Conference on Foundations of Software Technology and Theoretical Computer Science*, volume 8 of *Leibnitz International Proceedings in Informatics*, pages 400–411, Chennai, India, December 2010. Dagstuhl Publishing.

[111] Marcel-Paul Schützenberger. On context-free languages and push-down automata. *Information and Control*, 6(3):246–264, September 1963.

[112] Larry J. Stockmeyer and Albert R. Meyer. Word problems requiring exponential time: Preliminary report. In Alfred V. Aho, Allan Borodin, Robert L. Constable, Robert W. Floyd, Michael A. Harrison, Richard M. Karp, and H. Raymond Strong, editors, *Proceedings of the 5th Annual ACM Symposium on Theory of Computing*, pages 1–9, Austin, Texas, USA, April 1973. ACM.

[113] Róbert Szelepcsényi. The method of forced enumeration for nondeterministic automata. *Acta Informatica*, 26(3):279–284, November 1988.

[114] Andrzej Szepietowski. Closure properties of hyper-minimized automata. *RAIRO - Theoretical Informatics and Applications*, 45(4):459–466, November 2011.

[115] Robert Tarjan. Depth-first search and linear graph algorithms. *SIAM Journal on Computing*, 1(2):146–160, June 1972.

[116] Ken Thompson. Regular expression search algorithm. *Communications of the ACM*, 11(6):419–422, June 1968.

[117] Seinosuke Toda. *Computational Complexity of Counting Complexity Classes*. PhD Thesis, Tokyo Institute of Technology, Department of Computer Science, Tokyo, Japan, 1991.

[118] Boris A. Trakhtenbrot. Finite automata and the logic of monadic predicates. *Doklady Akademii Nauk SSSR*, 140:326–329, 1961.

[119] Boris A. Trakhtenbrot. A survey of russian approaches to perebor (brute-force search) algorithms. *Annals of the History of Computing*, 6(4):384–400, October 1984.

[120] Alan M. Turing. On computable numbers, with an application to the Entscheidungsproblem. *Proceedings of the London Mathematical Society*, s2-42(1):230–265, November 1936.

[121] Leslie G. Valiant. The complexity of enumeration and reliability problems. *SIAM Journal on Computing*, 8(3):410–421, August 1979.

[122] Leslie G. Valiant. The complexity of computing the permanent. *Theoretical Computer Science*, 8(2):189–201, 1979.

[123] Bruce W. Watson. *Taxonomies and Toolkits of Regular Language Algorithms*. PhD Thesis, Eindhoven University of Technology, Deparment of Mathematics and Computing Science, Eindhoven, The Netherlands, September 1995.

[124] Raymond T. Yeh. Structural equivalence of automata. *Mathematical Systems Theory*, 4(2):189–211, June 1970.

[125] Hsu-Chun Yen and Oscar H. Ibarra, editors. *Developments in Language Theory*, volume 7410 of *Lecture Notes in Computer Science*, Taipei, Taiwan, August 2012. Springer.

[126] Sheng Yu. Regular languages. In Rozenberg and Salomaa [107], chapter 2, pages 41–110.

Index

A^R, *see* reversal automaton,
    see dual biautomaton
$L(A)$, *see* accepted language
$L(G)$, *see* generated language
$L(q)$, *see* right language
$L(r)$, *see* described language
L^*, *see* Kleene star
L^+, 12
L^R, *see* reversal
L^i, 12
$L_A(q)$, *see* right language
L_q, *see* left language
$L_{A,q}$, *see* left language
Lw^{-1}, *see* derivative
$Q_A(E)$, 73
$\# \cdot C$, 64
$\mathrm{Ker}(A)$, *see* kernel
$\mathcal{P}(A)$, *see* power-set automaton
$\mathrm{Pre}(A)$, *see* preamble
$\Rightarrow_G$, *see* derivation
$\Rightarrow_G^*$, *see* derivation
$|w|$, *see* length
$|w|_a$, 11
$\approx_\ell$, *see* similar
$\cdot$, *see* concatenation
$\circlearrowleft(L)$, *see* cyclic shift
$\diamond$-property, *see* diamond-property
$\equiv$, *see* equivalence
$\equiv_L$, *see* syntactic congruence
λ, *see* empty word
$\overline{L}$, *see* complement
$\backslash$, *see* difference
$\sim$, *see* almost-equivalence
$\sim_E$, *see* E-equivalence

$\sim_k$
    for automata, *see* k-equivalence
    for languages, *see* k-equivalence
    for states, *see* k-similar
$\triangle$, *see* symmetric difference
w^R, *see* reversal
w^i, 12
$w^{-1}L$, *see* derivative

accept, 16, 93
accepted language, 16, 93
Aho, Alfred V., 8, 41, 42
Almeida, Marco, 71
almost-equivalence
    of automata, 29, 156
    of languages, 29, 156
    of states, 29, 156
almost-equivalence problem, 42
alphabet, 11, 92
alphabetic width, 133
Àlvarez, Carme, 64

Badr, Andrew, 8, 28–32, 38, 59, 156
Baier, Christel, 27
Berry, Gerard, 133
biautomaton
    deterministic, 94
    nondeterministic, 92
Birget, Jean-Camille, 52, 130
Björklund, Henrik, 45
Bonchi, Filippo, 27
both-sided derivative, *see* derivative
Boute, Raymond, 28